Introduction to Biology

Introduction to Biology

Melody Glover

Larsen & Keller
www.larsen-keller.com

Introduction to Biology
Melody Glover
ISBN: 978-1-64172-359-6 (Hardback)

© 2020 Larsen & Keller

⊟ Larsen & Keller

Published by Larsen and Keller Education,
5 Penn Plaza,
19th Floor,
New York, NY 10001, USA

Cataloging-in-Publication Data

Introduction to biology / Melody Glover.
    p. cm.
Includes bibliographical references and index.
ISBN 978-1-64172-359-6
1. Biology. 2. Life sciences. 3. Natural history. I. Glover, Melody.
QH307.2 .I58 2020
570--dc23

For more information regarding Larsen and Keller Education and its products, please visit the publisher's website www.larsen-keller.com

Table of Contents

Preface

This book aims to help a broader range of students by exploring a wide variety of significant topics related to this discipline. It will help students in achieving a higher level of understanding of the subject and excel in their respective fields. This book would not have been possible without the unwavered support of my senior professors who took out the time to provide me feedback and help me with the process. I would also like to thank my family for their patience and support.

Biology is a branch of science which deals with the study of life and living organisms. It observes the physical structure, molecular interactions, physiological mechanisms, evolution and development of organisms. It is a natural science that includes the study of the cell as a basic unit of life, genes as the basic unit of inheritance and evolution as the force that drives the creation and extinction of species. There are various branches of biology, such as anatomy, microbiology, botany, cell biology and genetics. Anatomy is the study of the structures of organisms and microbiology studies the microorganisms as well as their interaction with other living things. Botany is involved in the study of plants and cell biology is the study of cell and the molecular and chemical interactions that occur within living cells. Genetics is a branch of biology that examines and studies genes and heredity in organisms. This book provides comprehensive insights into the field of biology. Some of the diverse topics covered herein address the varied branches that fall under this category. Those in search of information to further their knowledge will be greatly assisted by this book.

A brief overview of the book contents is provided below:

Chapter – What is Biology?

The natural science which is involved in the study of living organisms as well as their physical structure, molecular interactions, evolution, development and chemical processes is called biology. This chapter will provide a brief introduction to the key aspects of biology as well as biological engineering.

Chapter – Branches of Biology

There are numerous branches of biology such as anatomy, biochemistry, biophysics, developmental biology, mathematical and theoretical biology, plant biology, and animal biology. The topics elaborated in this chapter will help in gaining a better perspective about these branches of biology as well as their subdivisions.

Chapter – Genetics: The Study of Genes

The branch of biology which focuses on the study of genes, genetic variation and heredity in organisms is termed as genetics. Some of the major areas of study within this field are regulation of gene expression, extranuclear inheritance and genetic linkage. This chapter discusses in detail these key focus areas related to genetics.

Chapter – Molecular Biology

The branch of biology which focuses on the molecular basis of biological activity between biomolecules in different systems of a cell is called molecular biology. Some of the diverse techniques used within this field are polymerase chain reactions, southern blotting and northern blotting. This chapter closely examines these key techniques of molecular biology to provide an extensive understanding of the subject.

Chapter – Systems Biology

The interdisciplinary field of study which focuses on complex interactions within biological systems through a holistic approach towards biological research is known as systems biology. It is involved in the computational and mathematical analysis and modeling of complex biological systems. All the diverse principles and models related to systems biology have been carefully analyzed in this chapter.

Melody Glover

1
What is Biology?

The natural science which is involved in the study of living organisms as well as their physical structure, molecular interactions, evolution, development and chemical processes is called biology. This chapter will provide a brief introduction to the key aspects of biology as well as biological engineering.

Biology is the "science of life." It is the study of living and once-living things, from submicroscopic structures in single-celled organisms to entire ecosystems with billions of interacting organisms; it further ranges in time focus from a single metabolic reaction inside a cell to the life history of one individual and on to the course of many species over eons of time. Biologists study the characteristics and behaviors of organisms, how species and individuals come into existence, and their interactions with each other and with the environment. The purview of biology extends from the origin of life to the fundamental nature of human beings and their relationship to all other forms of life.

Biology, or "life science," offers a window into fundamental principles shared by living organisms. These principles reveal a harmony and unity of the living world operating simultaneously among a great diversity of species and even in the midst of competition both between and within species for scarce resources. The overlying harmony is seen at each level, from within a cell to the level of systems in individuals (nervous, circulatory, respiratory, etc.), the immediate interactions of one organism with others, and on to the complex of organisms and interactions comprising an ecosystem with a multitude of ecological niches each supporting one species. Such harmony is manifested in many universally shared characteristics among living beings, including interdependence, a common carbon-based biochemistry, a widespread pattern of complementary polarities, sexual reproduction, and homeostasis.

As the science dealing with all life, biology encompasses a broad spectrum of academic fields that have often been viewed as independent disciplines. Among these are molecular biology, biochemistry, cell biology, physiology, anatomy, developmental biology, genetics, ecology, paleontology, and evolutionary biology. While competition among individuals expressing genetic variability has generally been identified as a key factor in evolutionary development, the pivotal roles of cooperation and long-term symbiosis or symbiogenesis in living systems have emerged in the late twentieth century as essential complementary focal points for understanding both the origin of species and the dynamics of biological systems.

Biology studies the unity and variety of life.

Principles of Biology

While biology is unlike physics in that it does not usually describe biological systems in terms of objects that exclusively obey immutable physical laws described by mathematics, it is nevertheless characterized by several major principles and concepts, which include: universality, evolution, interactions, diversity, and continuity.

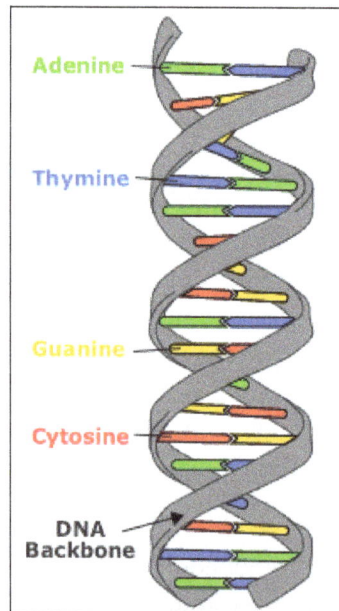
Schematic representation of DNA the primary genetic material.

Living organisms share many universal characteristics, including that they are composed of cells; pass on their heredity using a nearly universal genetic code; need energy from the environment to exist, grow, and reproduce; maintain their internal environment; and exhibit dual characteristics or complementary polarities. This are the common set of characteristics identified by biologists that distinguish living organisms from nonliving things.

With the exception of viruses, all organisms consist of cells, which are the basic units of life, being the smallest unit that can carry on all the processes of life, including maintenance, growth, and even self-repair. Some simple life forms, such as the paramecium, consist of a single cell throughout their life cycle and are called unicellular organisms. Multicellular organisms, such as a whale or tree, may have trillions of cells differentiated into many diverse types each performing a specific function.

All cells, in turn, are based on a carbon-based biochemistry, and all organisms pass on their heredity via genetic material based on nucleic acids such as DNA using a nearly universal genetic code. Every cell, no matter how simple or complex, utilizes nucleic acids for transmitting and storing the information needed for manufacturing proteins.

Every living being needs energy from the environment in order to exist, grow, and reproduce. Radiation from the sun is the main source of energy for life and is captured through photosynthesis, the biochemical process in which plants, algae, and some bacteria harness the energy of sunlight to produce food. Ultimately, nearly all living things depend on energy produced from photosynthesis for their nourishment, making it vital to life on Earth. There are also some bacteria that utilize the oxidation of inorganic compounds such as hydrogen sulfide or ferrous iron as an energy source. An organism that produces organic compounds from carbon dioxide as a carbon source, using either light or reactions of inorganic chemical compounds as a source of energy, is called an autotroph. Other organisms do not make their own food but depend directly or indirectly on autotrophs for their food. These are called heterotrophs.

In development, the theme of universal processes is also present. Living things grow and develop as they age. In most metazoan organisms the basic steps of the early embryo development share similar morphological stages and include similar genes.

All living organisms, whether unicellular or multicellular, exhibit homeostasis. Homeostasis is the property of an open system to regulate its internal environment so as to maintain a stable condition. Homeostasis can manifest itself at the cellular level through the maintenance of a stable internal acidity (pH); at the organismal level, warm-blooded animals maintain a constant internal body temperature; and at the level of the ecosystem, for example when atmospheric carbon dioxide levels rise, plants are theoretically able to grow healthier and thus remove more carbon dioxide from the atmosphere. Tissues and organs can also maintain homeostasis.

In addition, living beings share with all existent beings the quality of dual characteristics or complementary polarities. One common pair of dual characteristics is the quality of positivity and negativity: Just as sub-atomic particles have positive (electron) and negative (proton) elements that interrelate and form atoms, living beings commonly exhibit positive and negative characteristics. Most animals reproduce through relationships between male and female, and higher plants likewise have male and female elements, such as the (male) stamen and (female) pistil in flowering plants (angiosperms). Lower plants, fungi, some of the protists, and bacteria likewise exhibit reproductive variances, which are usually symbolized by + and - signs (rather than being called male and female), and referred to as "mating strains" or "reproductive types" or similar appellations.

Another more philosophical concept is the universal dual characteristic of within each organism of the invisible, internal character or nature and the visible aspects of matter, structure, and shape. For example, an animal will exhibit the internal aspects of life, instinct, and function of its cells, tissues, and organs, which relate with the visible shape made up by those cells, tissues, and organs.

Sexual reproduction is a trait that is almost universal among eukaryotes. Asexual reproduction is not uncommon among living organisms. In fact, it is widespread among fungi and bacteria, many insects reproduce in this manner, and some reptiles and amphibians. Nonetheless, with the exception of bacteria (prokaryotes), sexual reproduction is also seen in these same groups. (Some treat the unidirectional lateral transfer of genetic material in bacteria, between donors (+ mating type) and recipients (- mating type), as a type of sexual reproduction.) Evolutionary biologist and geneticist John Maynard Smith maintained that the perceived advantage for an individual organism to pass only its own entire genome to its offspring is so great that there must be an advantage by at least a factor of two to explain why nearly all animal species maintain a male sex.

Another characteristic of living things is that they take substances from the environment and organize them in complex hierarchical levels. For example, in multicellular organisms, cells are organized into tissues, tissues are organized into organs, and organs are organized into systems.

In addition, all living beings respond to the environment; that is, they react to a stimulus. A cockroach may respond to light by running for a dark place. When there is a complex set of response, it is called a behavior. For example, the migration of salmon is a behavioral response.

Evolution: A Common Organizing Principle Of Biology

A central, organizing concept in biology is that all life has descended from a common origin through a process of evolution. Indeed, eminent evolutionist Theodosius Dobzhansky has stated that "Nothing in biology makes sense except in the light of evolution." Evolution can be considered a unifying theme of biology because the concept of descent with modification helps to explain the common carbon-based biochemistry, the nearly universal genetic code, and the similarities and relationships among living organisms, as well as between organisms of the past with organisms today.

Evolutionary theory actually comprises several distinct components. Two of the major strands are the theory of descent with modification, which addresses the "pattern" of evolution, and the theory of natural selection, which addresses the "process" of evolution. Charles Darwin established evolution as a viable theory by marshaling and systematizing considerable evidence for the theory of descent with modification, including evidence from paleontology, classification, biogeography, morphology, and embryology. The mechanism that Darwin postulated, natural selection, aims to account for evolutionary changes at both the microevolutionary level (i.e., gene changes on the populational level) and the macroevolutionary level (i.e., major transitions between species and origination of new designs). Experimental tests and observations provide strong evidence for microevolutionary change directed by natural selection operating on heritable expressed variation, while evidence that natural selection directs macroevolution is limited to fossil evidence of some key transition sequences and extrapolation from evidences on the microevolutionary level. (Alfred Russel Wallace is commonly recognized as proposing the theory of natural selection at about the same time as Darwin.)

The evolutionary history of a species—which tells the characteristics of the various species from which it descended—together with its genealogical relationship to every other species is called its phylogeny. Widely varied approaches to biology generate information about phylogeny. These include the comparisons of DNA sequences conducted within molecular biology or genomics, and comparisons of fossils or other records of ancient organisms in paleontology. Biologists organize and analyze evolutionary relationships through various methods, including phylogenetics, phenetics, and cladistics. Major events in the evolution of life, as biologists currently understand them, are summarized on an evolutionary timeline.

Interactions: Harmony and Bi-level Functionality

Symbiosis between clownfish of the genus Amphiprion that dwell among the tentacles of tropical sea anemones.

The territorial fish protects the anemone from anemone-eating fish, and in turn the stinging tentacles of the anemone protect the clownfish from its predators.

Every living thing interacts with other organisms and its environment. One of the reasons that biological systems can be difficult to study is that there are so many different possible interactions with other organisms and the environment. A microscopic bacterium responding to a local gradient in sugar is as much responding to its environment as a lion is responding to its environment when it is searching for food in the African savanna. Within a particular species, behaviors can be cooperative, aggressive, parasitic, or symbiotic.

Matters become more complex still when two or more different species interact in an ecosystem, studies of which lie in the province of ecology. Analysis of ecosystems shows that a major factor in maintaining harmony and reducing competition is the tendency for each species to find and occupy a distinctive niche not occupied by other species.

Overlying the interactions of organisms is a sense of unity and harmony at each level of interaction. On the global level, for example, one can see the harmony between plant and animal life in terms of photosynthesis and respiration. Plants, through photosynthesis, use carbon dioxide and give off oxygen. While they also respire, plants' net input to the globe is considerably more oxygen than they consume (with algae in the ocean being a major source of planetary oxygen). Animals, on the other hand, consume oxygen and discharge carbon dioxide.

On the trophic level, the food web demonstrates harmony. Plants convert and store the sun's energy. These plants serve as food for herbivores, which in turn serve as food for carnivores, which are consumed by top carnivores. Top carnivores (and species at all other trophic levels), when dead, are broken down by decomposers such as bacteria, fungi, and some insects into minerals and humus in the soil, which is then used by plants.

On the level of individuals, the remarkable harmony among systems (nervous, circulatory, respiratory, endocrine, reproductive, skeletal, digestive, etc.) is a wonder to behold. Even within a cell, one sees remarkable examples of unity and harmony, such as when a cell provides a product to the body (such as a hormone) and receives oxygen and nourishment from the body. So remarkable is the harmony evident among organisms, and between organisms and the environment, that some have proposed a theory that the entire globe acts as if one, giant, functioning organism (the Gaia theory). According to well-known biologist Lynn Margulis and science writer Dorion Sagan (Microcosmos, 1997), even evolution is tied to cooperation and mutual dependence among organisms: "Life did not take over the globe by combat, but by networking."

An underlying explanation for such observed harmony is the concept of bi-level functionality, the view that every entity exists in an integral relation with other entities in ways that permit an individual entity to advance its own multiplication, development, self-preservation, and self-strengthening (a function for the individual) while at the same time contribute toward maintaining or developing the larger whole (a function for the whole). These functions are not independent but interdependent. The individual's own success allows it to contribute to the whole, and while the individual contributes something of value to the larger entity, assisting the larger entity in advancing its own function, the larger entity likewise provides the environment for the success of the individual.

For example, in the cells of a multicellular organism, each cell provides a useful function for the body as a whole. A cell's function may be to convert sugar to ADP energy, attack foreign invaders, or produce hormones. A cell in the epithelial tissue of the stomach may secrete the enzyme pepsin to help with digestion. The cell's function of providing pepsin to the body is harmonized with the body's needs for maintenance, development, and reproduction. The body, on the other hand, supports the individual cell and its function by providing food, oxygen, and other necessary materials, and by transporting away the toxic waste materials. Each cell actually depends on the other cells in the body to perform their functions and thus keep the body in proper functioning order. Likewise, a particular taxonomic group (taxa) not only advances its own survival and reproduction, but also provides a function for the ecosystems of which it is part, such as the ocelot species helping to regulate prey populations and thus help ecosystems to maintain balance. An ecosystem provides an environment for the success of this taxonomic group and thus its contribution to the ecosystem. In essence, this explanation holds that while animals and plants may seem to struggle against one another for existence, in reality they do not. Rather, they all contribute to the whole, in harmony.

Human beings, the most complex of all biological organisms, likewise live in a biosphere that is all interrelated and is necessary for physical life. Thus, it becomes essential that human beings, as the most powerful of all life forms and in many ways an encapsulation of the whole (a "microcosm of creation" according to a theological perspective), understand and care for the environment. In religious terms, this is sometimes referred to as the "third blessing," the role of humankind to love and care for creation. The science of biology is central to this process.

The science of physics offers complementary rationales both for explaining evolutionary development and also for urging humans to love and care for the biosphere. This striking advance in physics arises through the extension of the second law of thermodynamics to apply to "open" systems, which include all forms of life. The extended second law states simply that natural processes in open systems tend to dissipate order as rapidly as possible. From this perspective, evolution of life's successively more ordered and complex systems occurs because the greater a system's order and complexity, the greater its capacity to dissipate order. Human beings, as the planet's dominant and most complex species, face a thermodynamic imperative to apply themselves toward establishing an even greater level of order and dynamic complexity on the planet. Achieving such greater order would likely require that humans learn to live together in peace while living in synergy with the biosphere.

Diversity: The Variety of Living Organisms

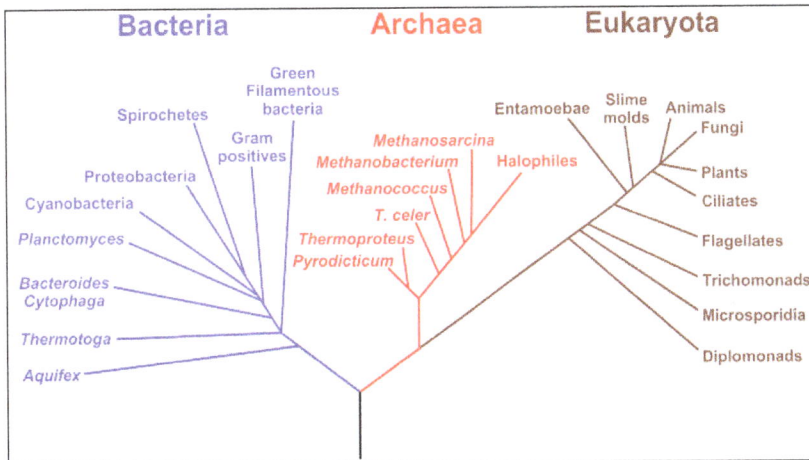

A phylogenetic tree of all living things, based on rRNA gene data, showing the separation of the three domains bacteria, archaea, and eukaryotes as described initially by Carl Woese.

Trees constructed with other genes are generally similar, although they may place some early branching groups very differently. The exact relationships of the three domains are still being debated.

Despite the underlying unity, life exhibits an astonishing wide diversity in morphology, behavior, and life histories. In order to grapple with this diversity, biologists, following a conventional western scientific approach and historically unaware of the profound interdependence of all life on the planet, attempt to classify all living things. This scientific classification should reflect the evolutionary trees (phylogenetic trees) of the different organisms. Such classifications are the province of the disciplines of systematics and taxonomy. Taxonomy puts organisms in groups called taxa, while systematics seeks their relationships.

Until the nineteenth century, living organisms were generally divided into two kingdoms: animal and plant, or the Animalia and the Plantae. As evidence accumulated that these divisions were insufficient to express the diversity of life, schemes with three, four, or more kingdoms were proposed.

A popular scheme, developed in 1969 by Robert Whitaker, delineates living organisms into five kingdoms:

Monera — Protista — Fungi — Plantae —Animalia.

In the six-kingdom classification, the six top-level groupings (kingdoms) are:

Archaebacteria, Monera (the bacteria and cyanobacteria), Protista, Fungi, Plantae, and Animalia.

These schemes coexist with another scheme that divides living organisms into the two main divisions of prokaryote (cells that lack a nucleus: bacteria, etc.) and eukaryote (cells that have a nucleus and membrane-bound organelles: animals, plants, fungi, and protists).

In 1990, another scheme, a three-domain system, was introduced by Carl Woese and has became very popular (with the "domain" a classification level higher than kingdom):

> Archaea (originally Archaebacteria) — Bacteria (originally Eubacteria) — Eukaryota (or Eucarya).

The three-domain system is a biological classification that emphasizes his separation of prokaryotes into two groups, the Bacteria and the Archaea (originally called Eubacteria and Archaebacteria). When recent work revealed that what were once called "prokaryotes" are far more diverse than suspected, the prokaryotes were divided into the two domains of the Bacteria and the Archaea, which are considered to be as different from each other as either is from the eukaryotes. Woese argued based on differences in 16S ribosomal RNA genes that these two groups and the eukaryotes each arose separately from an ancestral progenote with poorly developed genetic machinery. To reflect these primary lines of descent, he treated each as a domain, divided into several different kingdoms. The groups were also renamed the Bacteria, Archaea, and Eukaryota, further emphasizing the separate identity of the two prokaryote groups.

There is also a series of intracellular "parasites" that are progressively less alive in terms of being metabolically active:

> Viruses — Viroids — Prions

Continuity: The Common Descent of Life

A group of organisms is said to have common descent if they have a common ancestor. All existing organisms on Earth are descended from a common ancestor or ancestral gene pool. This "last universal common ancestor," that is, the most recent common ancestor of all organisms, is believed to have appeared about 3.5 billion years ago.

The notion that "all life is from an egg" is a foundational concept of modern biology, it means that there has been an unbroken continuity of life from the initial origin of life to the present time. Up into the nineteenth century it was commonly believed that life forms can appear spontaneously under certain conditions (abiogenesis).

The universality of the genetic code is generally regarded by biologists as strong support of the theory of universal common descent (UCD) for all bacteria, archaea, and eukaryotes.

Scope of Biology

Academic Disciplines

Biologists study life over a wide range of scales: Life is studied at the atomic and molecular scale in molecular biology, biochemistry, and molecular genetics. At the level of the cell, life is studied in cell biology, and at multicellular scales, it is examined in physiology, anatomy, and histology. Developmental biology involves study of life at the level of the development or ontogeny of an individual organism.

Moving up the scale toward more than one organism, genetics considers how heredity works between parent and offspring. Ethology considers group behavior of organisms. Population genetics looks at the level of an entire population, and systematics considers the multi-species scale of lineages. Interdependent populations and their habitats are examined in ecology.

Two broad disciplines within biology are botany, the study of plants, and zoology, the study of animals. Paleontology is inquiry into the developing history of life on earth, based on working with fossils, and includes the main subfields of paleobotany, paleozoology, and micropaleontology. Changes over time, whether within populations (microevolution) or involving either speciation or the introduction of major designs (macroevolution), is part of the field of inquiry of evolutionary biology. A speculative new field is astrobiology (or xenobiology) which examines the possibility of life beyond the Earth.

Biology has become such a vast research enterprise that it is not generally studied as a single discipline, but as a number of clustered sub-disciplines. Four broad groupings are considered here. The first broad group consists of disciplines that study the basic structures of living systems: cells, genes, and so forth; a second grouping considers the operation of these structures at the level of tissues, organs and bodies; a third grouping considers organisms and their histories; and a final constellation of disciplines focuses on the interactions. It is important to note, however, that these groupings are a simplified description of biological research. In reality, the boundaries between disciplines are very fluid and most disciplines borrow techniques from each other frequently. For example, evolutionary biology leans heavily on techniques from molecular biology to determine DNA sequences that assist in understanding the genetic variation of a population; and physiology borrows extensively from cell biology in describing the function of organ systems.

Ethical Aspects

As in all sciences, biological disciplines are best pursued by persons committed to high ethical standards, maintaining the highest integrity and following a good research methodology. Data should be interpreted honestly, and results that do not fit one's preconceived biases should not be discarded or ignored in favor of data that fits one's prejudices. A biologist who puts her or his own well-being first (money, popularity, position, etc.), runs the risk of faulty or even fraudulent research. But even well-meaning biologists have gone off course in trying to fit research findings to personal biases

Also overlying work in many biological fields is the more specific concept of bioethics. This is the discipline dealing with the ethical implications of biological research and its applications. Aspects of biology raising issues of bioethics include cloning, genetic engineering, population control, medical research on animals, creation of biological weapons, and so forth.

Structure of Life

Molecular biology is the study of biology at the molecular level. The field overlaps with other areas of biology, particularly genetics and biochemistry. Molecular biology chiefly concerns itself with understanding the interactions between the various systems of a cell, especially by mapping the interactions between DNA, RNA, and protein synthesis and learning how these interactions are regulated.

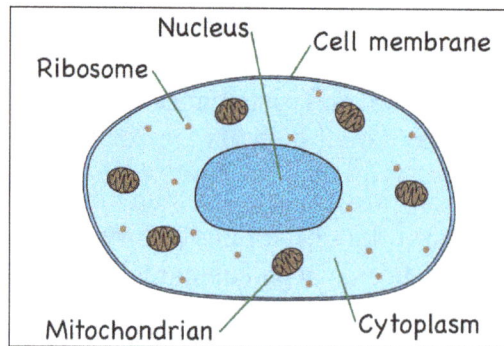
Schematic of typical animal cell.

Cell biology studies the physiological properties of cells, as well as their behaviors, interactions, and environment; this is done both on a microscopic and molecular level. Cell biology researches both single-celled organisms like bacteria and specialized cells in multicellular organisms like humans.

Understanding the composition of cells and how cells work is fundamental to all of the biological sciences. Appreciating the similarities and differences between cell types is particularly important to the fields of cell and molecular biology. These fundamental similarities and differences provide a unifying theme, allowing the principles learned from studying one cell type to be extrapolated and generalized to other cell types.

Genetics is the science of genes, heredity, and the variation of organisms. In modern research, genetics provides important tools in the investigation of the function of a particular gene (e.g., analysis of genetic interactions). Within organisms, genetic information generally is carried in chromosomes, where it is represented in the chemical structure of particular DNA molecules.

Genes encode the information necessary for synthesizing proteins, which in turn play a large role in influencing the final phenotype of the organism, although in many instances do not completely determine it.

Developmental biology studies the process by which organisms grow and develop. Originating in embryology, today, developmental biology studies the genetic control of cell growth, differentiation, and "morphogenesis," which is the process that gives rise to tissues, organs, and anatomy. Model organisms for developmental biology include the round worm Caenorhabditis elegans, the fruit fly Drosophila melanogaster, the zebrafish Brachydanio rerio, the mouse Mus musculus, and the small flowering mustard plant Arabidopsis thaliana.

Physiology of Organisms

Physiology studies the mechanical, physical, and biochemical processes of living organisms, by attempting to understand how all the structures function as a whole. The theme of "structure to function" is central to biology.

Physiological studies have traditionally been divided into plant physiology and animal physiology, but the principles of physiology are universal, regardless of the particular organism being studied. For example, what is learned about the physiology of yeast cells can also apply to other cells. The field of animal physiology extends the tools and methods of human physiology to non-human animal species. Plant physiology also borrows techniques from both fields.

Anatomy is an important part of physiology and considers how organ systems in animals such as the nervous, immune, endocrine, respiratory, and circulatory systems function and interact. The study of these systems is shared with the medically oriented disciplines of neurology, immunology, and the like. The field of health science deals with both human and animal health.

Diversity and Evolution of Organisms

In population genetics the evolution of a population of organisms is sometimes depicted as if traveling on a fitness landscape. The arrows indicate the preferred flow of a population on the landscape, and the points A, B, and C are local optima. The red ball indicates a population that moves from a very low fitness value to the top of a peak.

Evolutionary biology is concerned with the origin and descent of species, and their change over time, i.e., their evolution. Evolutionary biology is an inclusive field because it includes scientists from many traditional taxonomically oriented disciplines. For example, it generally includes scientists who may have a specialist training in particular organisms such as mammalogy, ornithology, or herpetology, but uses those organisms as systems to answer general questions in evolution. It also generally includes paleontologists who use fossils to answer questions about the mode and tempo of evolution, as well as theoreticians in areas such as population genetics and evolutionary theory. In the 1990s, developmental biology made a re-entry into evolutionary biology from its initial exclusion from the modern synthesis through the study of evolutionary developmental biology. Related fields which are often considered part of evolutionary biology are phylogenetics, systematics, and taxonomy.

The two major traditional taxonomically oriented disciplines are botany and zoology. Botany is the scientific study of plants. It covers a wide range of scientific disciplines that study the growth, reproduction, metabolism, development, diseases, and evolution of plant life. Zoology is the discipline that involves the study of animals, which includes the physiology of animals studied under various fields, including anatomy and embryology. The common genetic and developmental mechanisms of animals and plants is studied in molecular biology, molecular genetics, and developmental biology. The ecology of animals is covered under behavioral ecology and other fields.

Classification of Life

The dominant classification system is called Linnaean taxonomy, which includes ranks and binomial nomenclature. How organisms are named is governed by international agreements such as the International Code of Botanical Nomenclature (ICBN), the International Code of Zoological Nomenclature (ICZN), and the International Code of Nomenclature of Bacteria (ICNB). A fourth Draft BioCode was published in 1997 in an attempt to standardize naming in the three areas, but it has not yet been formally adopted. The International Code of Virus Classification and Nomenclature (ICVCN) remains outside the BioCode.

Interactions of Organisms

Ecology studies the distribution and abundance of living organisms, and the interactions between organisms and their environment. The environment of an organism includes both its habitat, which can be described as the sum of local abiotic factors like climate and geology, as well as the other organisms which share its habitat. Ecological systems are studied at several different levels—from individuals and populations to ecosystems and the biosphere level. Ecology is a multi-disciplinary science, drawing on many other branches of science.

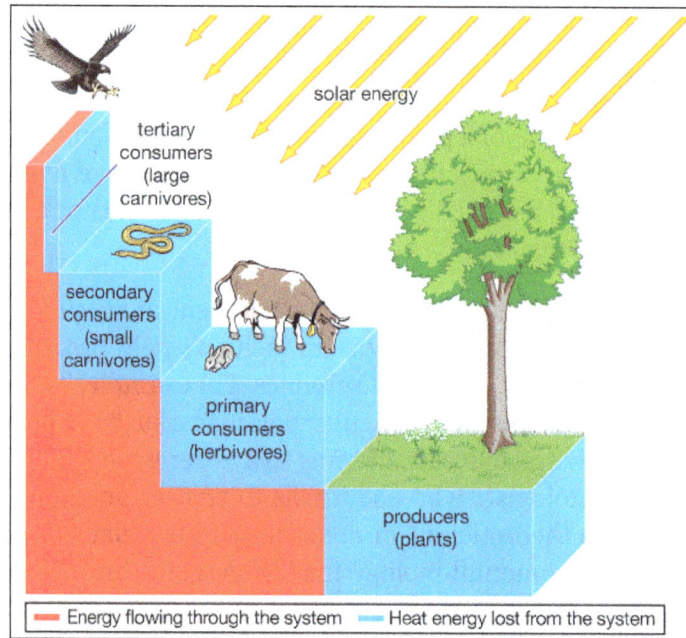

A food web, a generalization of the food chain, depicting the complex interrelationships among organisms in an ecosystem.

Ethology studies animal behavior (particularly of social animals such as primates and canids), and is sometimes considered as a branch of zoology. Ethologists have been particularly concerned with the evolution of behavior and the understanding of behavior in terms of evolutionary thought. In one sense, the first modern ethologist was Charles Darwin, whose book The Expression of the Emotions in Animals and Men influenced many ethologists.

Cell Theory

Cell Theory is one of the basic principles of biology. Credit for the formulation of this theory is given to German scientists Theodor Schwann, Matthias Schleiden, and Rudolph Virchow.

The Cell Theory States

- All living organisms are composed of cells. They may be unicellular or multicellular.

- The cell is the basic unit of life.

- Cells arise from pre-existing cells. (They are not derived from spontaneous generation.)

The modern version of the cell theory includes the ideas that:

- Energy flow occurs within cells.

- Heredity information (DNA) is passed on from cell to cell.

- All cells have the same basic chemical composition.

In addition to the cell theory, the gene theory, evolution, homeostasis, and the laws of thermodynamics form the basic principles that are the foundation for the study of life.

Cell Basics

All living organisms in the kingdoms of life are composed of and depend on cells to function normally. Not all cells, however, are alike. There are two primary types of cells: eukaryotic and prokaryotic cells. Examples of eukaryotic cells include animal cells, plant cells, and fungal cells. Prokaryotic cells include bacteria and archaeans.

Cells contain organelles, or tiny cellular structures, that carry out specific functions necessary for normal cellular operation. Cells also contain DNA (deoxyribonucleic acid) and RNA (ribonucleic acid), the genetic information necessary for directing cellular activities.

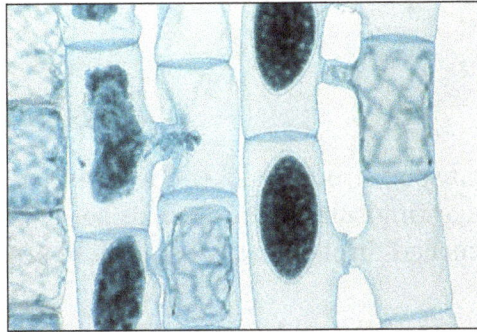

Eukaryotic cells grow and reproduce through a complex sequence of events called the cell cycle. At the end of the cycle, cells will divide either through the processes of mitosis or meiosis. Somatic cells replicate through mitosis and sex cells reproduce via meiosis. Prokaryotic cells reproduce commonly through a type of asexual reproduction called binary fission. Higher organisms are also capable of asexual reproduction. Plants, algae, and fungi reproduce through the formation of reproductive cells called spores. Animal organisms can reproduce asexually through processes such as budding, fragmentation, regeneration, and parthenogenesis.

Cell Processes: Cellular Respiration and Photosynthesis

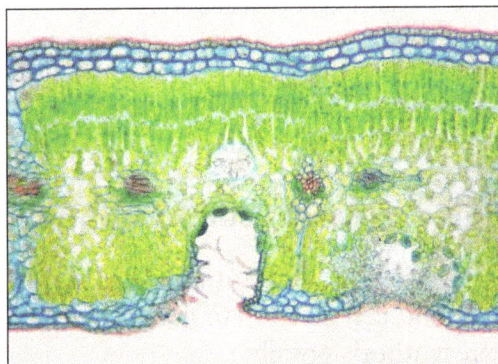

Cells perform a number of important processes that are necessary for the survival of an organism. Cells undergo the complex process of cellular respiration in order to obtain energy stored in the nutrients consumed. Photosynthetic organisms including plants, algae, and cyanobacteria are capable of photosynthesis. In photosynthesis, light energy from the sun is converted to glucose. Glucose is the energy source used by photosynthetic organisms and other organisms that consume photosynthetic organisms.

Cell Processes: Endocytosis and Exocytosis

Cells also perform the active transport processes of endocytosis and exocytosis. Endocytosis is the process of internalizing and digesting substances, such as seen with macrophages and bacteria. The digested substances are expelled through exocytosis. These processes also allow for molecule transportation between cells.

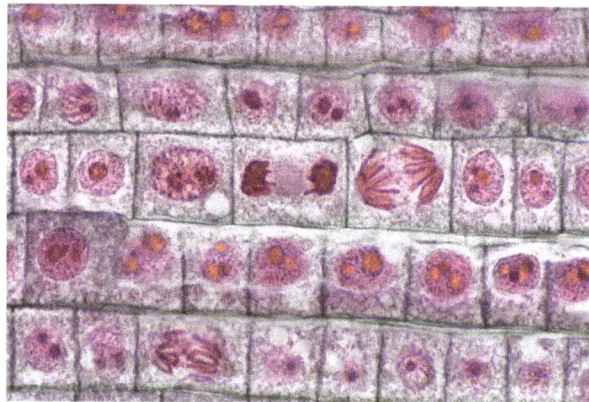

Cell migration is a process that is vital for the development of tissues and organs. Cell movement is also required for mitosis and cytokinesis to occur. Cell migration is made possible by interactions between motor enzymes and cytoskeleton microtubules.

Cell Processes: DNA Replication and Protein Synthesis

The cell process of DNA replication is an important function that is needed for several processes including chromosome synthesis and cell division to occur. DNA transcription and RNA translation make the process of protein synthesis possible.

Biological Engineering

Biological Engineering is an interdisciplinary area focusing on the application of engineering principles to analyze biological systems and to solve problems in the interfacing of such systems plant, animal or microbial with human-designed machines, structures, processes and instrumentation. The biological revolution continues to mature and impact all of us. Human-based gene manipulation affects nearly all North American food supplies. Plants and animals are already being defined on a molecular basis. Living organisms can now be analyzed, measured and "engineered" as never before. Designer "bugs" are being produced to enhance biological processes. These changes continue to redefine our research and graduate programs that continue to emphasize biological, environmental and food and fiber engineering. Our connections to agriculture and food systems remain, but modern agriculture is greatly influenced by biotechnology, and our connections to agriculture reflect this fact. A basic goal is to design technology that operates in harmony with the biology of living systems. In many cases, currently available knowledge is inadequate to support engineering design of food and biological processes. Hence, greater fundamental knowledge of biology and its potential applications are also of concern to the biological engineer.

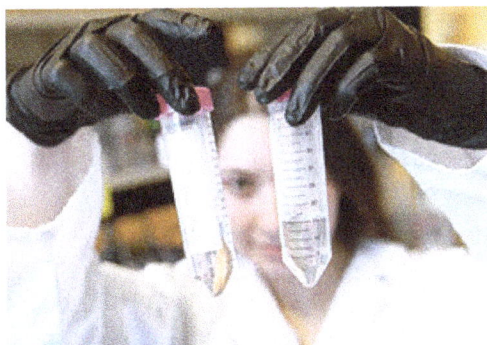

Department Research Areas Include

Molecular Bioengineering

Biosensors, bioassays and microfluidic lab-on-a-chip systems will be developed for the detection of pathogenic organisms, toxins, and clinically relevant markers. Applications of the sensors will be toward clinical diagnostics, food safety, environmental protection or biosecurity.

Nucleic Acid Engineering

Work with engineering DNA into a nanomaterial for real world applications including drug (DNA/siRNA/cell) delivery, molecular sensing, cell-free protein production, protein engineering and nanoparticle-based photonic/optoelectronic/photovoltaic devices.

Physiological Engineering

Experiment with measurement and modeling of physiological functions in animals and plants. A broad range of projects are possible and can involve physiological functions at the cellular level as well as larger more complex systems.

Microbial Fuel Cells

Microbial fuel cells use bacterial cells for the generation of bioelectricity from waste products. Projects within this area include gene expression studies using Bioconductor in the R- project language, development of a portable potentiostat to study bacterial interactions in situ, studying bacterial interactions, and development of a microfluidic bioreactor.

Bioenergetics and Stress Factors

Bioenergetics involves development of mechanistic models to predict energy budget of endotherms for virtually any thermal conditions. Stress factors in livestock involves time series analyses of thermal data (temperature, relative humidity, wind speed and solar radiation), physiological responses data (internal body temperature, respiration rate, sweating rate, etc) and physical and optical properties of hair coat (fur layer) in defining stress and stress levels of livestock in hot and dry, and hot and humid environments.

Energy Systems Engineering

A systematic approach to future energy needs. Projects within this area include: Developing and validating system models of material, energy and monetary flows for cellulosic and corn ethanol and application of models to assess system energetics, economics, and carbon balance; Developing frameworks for integration of uncertain wind resources into existing electric power grids through the use of optimization in conjunction with simulations on a simplified power system.

Controlled-environment Agriculture

Developing energy-efficient technologies to support year-round, local, vegetable production and production of high value chemicals such as pharmaceuticals from plants in closed environments.

Soil and Water Engineering

Developing, testing, and designing water quality protection strategies and novel approaches to monitoring hydrological systems. Additional focus on sustainable water development throughout the world.

2
Branches of Biology

There are numerous branches of biology such as anatomy, biochemistry, biophysics, developmental biology, mathematical and theoretical biology, plant biology, and animal biology. The topics elaborated in this chapter will help in gaining a better perspective about these branches of biology as well as their subdivisions.

Anatomy

Anatomy is a field in the biological sciences concerned with the identification and description of the body structures of living things. Gross anatomy involves the study of major body structures by dissection and observation and in its narrowest sense is concerned only with the human body. "Gross anatomy" customarily refers to the study of those body structures large enough to be examined without the help of magnifying devices, while microscopic anatomy is concerned with the study of structural units small enough to be seen only with a light microscope. Dissection is basic to all anatomical research. The earliest record of its use was made by the Greeks, and Theophrastus called dissection "anatomy," from ana temnein, meaning "to cut up."Comparative anatomy, the other major subdivision of the field, compares similar body structures in different species of animals in order to understand the adaptive changes they have undergone in the course of evolution.

Gross Anatomy

This ancient discipline reached its culmination between 1500 and 1850, by which time its subject matter was firmly established. None of the world's oldest civilizations dissected a human body, which most people regarded with superstitious awe and associated with the spirit of the departed soul. Beliefs in life after death and a disquieting uncertainty concerning the possibility of bodily resurrection further inhibited systematic study. Nevertheless, knowledge of the body was acquired by treating wounds, aiding in childbirth, and setting broken limbs. The field remained speculative rather than descriptive, though, until the achievements of the Alexandrian medical school and its foremost figure, Herophilus (flourished 300 BCE), who dissected human cadavers and thus gave anatomy a considerable factual basis for the first time. Herophilus made many important discoveries and was followed by his younger contemporary Erasistratus, who is sometimes regarded as the founder of physiology. In the 2nd century CE, Greek physician Galen assembled and arranged all the discoveries of the Greek anatomists, including with them his own concepts of physiology and his discoveries in experimental medicine. The many books Galen wrote became the unquestioned authority for anatomy and medicine in Europe because they were the only ancient Greek anatomical texts that survived the Dark Ages in the form of Arabic (and then Latin) translations.

Owing to church prohibitions against dissection, European medicine in the Middle Ages relied upon Galen's mixture of fact and fancy rather than on direct observation for its anatomical knowledge, though some dissections were authorized for teaching purposes. In the early 16th century, the artist Leonardo da Vinci undertook his own dissections, and his beautiful and accurate anatomical drawings cleared the way for Flemish physician Andreas Vesalius to "restore" the science of anatomy with his monumental De humani corporis fabrica libri septem (1543; "The Seven Books on the Structure of the Human Body"), which was the first comprehensive and illustrated textbook of anatomy. As a professor at the University of Padua, Vesalius encouraged younger scientists to accept traditional anatomy only after verifying it themselves, and this more critical and questioning attitude broke Galen's authority and placed anatomy on a firm foundation of observed fact and demonstration.

From Vesalius's exact descriptions of the skeleton, muscles, blood vessels, nervous system, and digestive tract, his successors in Padua progressed to studies of the digestive glands and the urinary and reproductive systems. Hieronymus Fabricius, Gabriello Fallopius, and Bartolomeo Eustachio were among the most important Italian anatomists, and their detailed studies led to fundamental progress in the related field of physiology. William Harvey's discovery of the circulation of the blood, for instance, was based partly on Fabricius's detailed descriptions of the venous valves.

Microscopic Anatomy

The new application of magnifying glasses and compound microscopes to biological studies in the second half of the 17th century was the most important factor in the subsequent development of anatomical research. Primitive early microscopes enabled Marcello Malpighi to discover the system of tiny capillaries connecting the arterial and venous networks, Robert Hooke to first observe the small compartments in plants that he called "cells," and Antonie van Leeuwenhoek to observe muscle fibres and spermatozoa. Thenceforth attention gradually shifted from the identification and understanding of bodily structures visible to the naked eye to those of microscopic size.

The use of the microscope in discovering minute, previously unknown features was pursued on a more systematic basis in the 18th century, but progress tended to be slow until technical improvements in the compound microscope itself, beginning in the 1830s with the gradual development of achromatic lenses, greatly increased that instrument's resolving power. These technical advances enabled Matthias Jakob Schleiden and Theodor Schwann to recognize in 1838–39 that the cell is the fundamental unit of organization in all living things. The need for thinner, more transparent tissue specimens for study under the light microscope stimulated the development of improved methods of dissection, notably machines called microtomes that can slice specimens into extremely thin sections. In order to better distinguish the detail in these sections, synthetic dyes were used to stain tissues with different colours. Thin sections and staining had become standard tools for microscopic anatomists by the late 19th century. The field of cytology, which is the study of cells, and that of histology, which is the study of tissue organization from the cellular level up, both arose in the 19th century with the data and techniques of microscopic anatomy as their basis.

In the 20th century anatomists tended to scrutinize tinier and tinier units of structure as new technologies enabled them to discern details far beyond the limits of resolution of light microscopes. These advances were made possible by the electron microscope, which stimulated an enormous amount of research on subcellular structures beginning in the 1950s and became the prime tool of anatomical research. About the same time, the use of X-ray diffraction for studying the structures of many types of molecules present in living things gave rise to the new subspecialty of molecular anatomy.

Anatomical Nomenclature

Scientific names for the parts and structures of the human body are usually in Latin; for example, the name musculus biceps brachii denotes the biceps muscle of the upper arm. Some such names were bequeathed to Europe by ancient Greek and Roman writers, and many more were coined by European anatomists from the 16th century on. Expanding medical knowledge meant the discovery of many bodily structures and tissues, but there was no uniformity of nomenclature, and thousands of new names were added as medical writers followed their own fancies, usually expressing them in a Latin form.

Comparative Anatomy

Comparative anatomy is the study of similarities and differences in the anatomy of different species. It is closely related to evolutionary biology and phylogeny (the evolution of species).

The science began in the classical era, continuing in Early Modern times with work by Pierre Belon who noted the similarities of the skeletons of birds and humans.

Comparative anatomy has provided evidence of common descent, and has assisted in the classification of animals.

Concepts

Two major concepts of comparative anatomy are:

1. Homologous structures - Structures (body parts/anatomy) which are similar in different species because the species have common descent and have evolved, usually divergently, from a shared ancestor. They may or may not perform the same function. An example is the forelimb structure shared by cats and whales.

2. Analogous structures - Structures similar in different organisms because, in convergent evolution, they evolved in a similar environment, rather than were inherited from a recent common ancestor. They usually serve the same or similar purposes. An example is the streamlined torpedo body shape of porpoises and sharks. So even though they evolved from different ancestors, porpoises and sharks developed analogous structures as a result of their evolution in the same aquatic environment. This is known as a homoplasy.

Uses

Comparative anatomy has long served as evidence for evolution, now joined in that role by comparative genomics; it indicates that organisms share a common ancestor. It also assists scientists in classifying organisms based on similar characteristics of their anatomical structures. A common example of comparative anatomy is the similar bone structures in forelimbs of cats, whales, bats, and humans. All of these appendages consist of the same basic parts; yet, they serve completely different functions. The skeletal parts which form a structure used for swimming, such as a fin, would not be ideal to form a wing, which is better-suited for flight. One explanation for the fore-limbs' similar composition is descent with modification. Through random mutations and natural selection, each organism's anatomical structures gradually adapted to suit their respective habi-tats. The rules for development of *special* characteristics which differ significantly from general homology were listed by Karl Ernst von Baer as the laws now named after him.

Surface Anatomy

Surface anatomy (also called superficial anatomy and visual anatomy) is the study of the external features of the body of an animal. In birds this is termed topography. Surface anatomy deals with anatomical features that can be studied by sight, without dissection. As such, it is a branch of gross anatomy, along with endoscopic and radiological anatomy. Surface anatomy is a descriptive sci-ence. In particular, in the case of human surface anatomy, these are the form and proportions of the human body and the surface landmarks which correspond to deeper structures hidden from view, both in static pose and in motion.

In addition, the science of surface anatomy includes the theories and systems of body proportions and related artistic canons. The study of surface anatomy is the basis for depicting the human body in classical art.

Some pseudo-sciences such as physiognomy, phrenology and palmistry rely on surface anatomy.

Human Surface Anatomy

Surface Anatomy of the Thorax

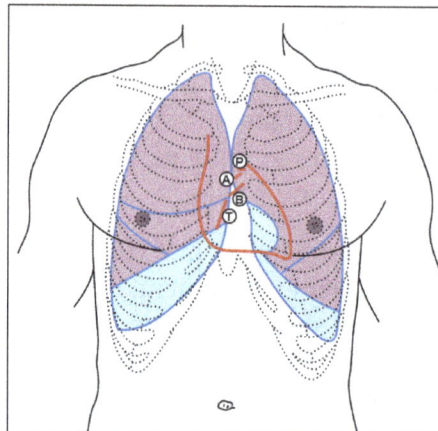

In figure, front of thorax, showing surface relations of bones, lungs (purple), pleura (blue), and heart (red outline). Heart valves are labeled with "B", "T", "A", and "P". First heart sound: caused by atrioventricular valves - Bicuspid/Mitral (B) and Tricuspid (T). Second heart sound caused by semilunar valves -- Aortic (A) and Pulmonary/Pulmonic (P).

Knowledge of the surface anatomy of the thorax (chest) is particularly important because it is one of the areas most frequently subjected to physical examination, like auscultation and percussion.

In cardiology, Erb's point refers to the third intercostal space on the left sternal border where S2 heart sound is best auscultated. Some sources include the fourth left interspace.

Human female breasts are located on the chest wall, most frequently between the second and sixth rib.

Anatomical Landmarks

On the trunk of the body in the thoracic area, the shoulder in general is the acromial, while the curve of the shoulder is the deltoid. The back as a general area is the dorsum or dorsal area, and the lower back as the limbus or lumbar region. The shoulderblades are the scapular area and the breastbone is the sternal region. The abdominal area is the region between the chest and the pelvis. The breast is called the mamma or mammary, the armpit as the axilla and axillary, and the navel as the umbilicus and umbilical. The pelvis is the lower torso, between the abdomen and the thighs. The groin, where the thigh joins the trunk, are the inguen and inguinal area. The entire arm is referred to as the brachium and brachial, the front of the elbow as the antecubitis and antecubital, the back of the elbow as the olecranon or olecranal, the forearm as the antebrachium and antebrachial, the wrist as the carpus and carpal area, the hand as the manus and manual, the palm as the palma and palmar, the thumb as the pollex, and the fingers as the digits, phalanges, and phalangeal. The buttocks are the gluteus or gluteal region and the pubic area is the pubis.

Anatomists divide the lower limb into the thigh (the part of the limb between the hip and the knee) and the leg (which refers only to the area of the limb between the knee and the ankle). The thigh is the femur and the femoral region. The kneecap is the patella and patellar while the back of the knee is the popliteus and popliteal area. The leg (between the knee and the ankle) is the crus and crural area, the lateral aspect of the leg is the peroneal area, and the calf is the sura and sural region. The ankle is the tarsus and tarsal, and the heel is the calcaneus or calcaneal. The foot is the pes and pedal region, and the sole of the foot the planta and plantar. As with the fingers, the toes are also called the digits, phalanges, and phalangeal area. The big toe is referred to as the hallux.

Biochemistry

Biochemistry is the study of chemical processes within and relating to living organisms. Biochemical processes give rise to the complexity of life.

A sub-discipline of both biology and chemistry, biochemistry can be divided in three fields; molecular genetics, protein science and metabolism. Over the last decades of the 20th century, biochemistry has through these three disciplines become successful at explaining living processes. Almost all areas of the life sciences are being uncovered and developed by biochemical methodology and research. Biochemistry and Molecular Biology focus on understanding how biological molecules give rise to the processes that occur within living cells and between cells, which in turn relates greatly to the study and understanding of tissues, organs, and organism structure and function.

Biochemistry is closely related to molecular biology, the study of the molecular mechanisms of biological phenomena.

Much of biochemistry deals with the structures, functions and interactions of biological macro-molecules, such as proteins, nucleic acids, carbohydrates and lipids, which provide the structure of cells and perform many of the functions associated with life. The chemistry of the cell also depends on the reactions of smaller molecules and ions. These can be inorganic, for example water and metal ions, or organic, for example the amino acids, which are used to synthesize proteins. The mechanisms by which cells harness energy from their environment via chemical reactions are known as metabolism. The findings of biochemistry are applied primarily in medicine, nutrition, and agriculture. In medicine, biochemists investigate the causes and cures of diseases. In nutrition, they study how to maintain health wellness and study the effects of nutritional deficiencies. In agriculture, biochemists investigate soil and fertilizers, and try to discover ways to improve crop cultivation, crop storage and pest control.

Gerty Cori and Carl Cori jointly won the Nobel Prize in 1947
for their discovery of the Cori cycle at RPMI.

At its broadest definition, biochemistry can be seen as a study of the components and composition of living things and how they come together to become life, in this sense the history of biochemistry may therefore go back as far as the ancient Greeks. However, biochemistry as a specific scientific discipline has its beginning sometime in the 19th century, or a little earlier, depending on which aspect of biochemistry is being focused on. Some argued that the beginning of biochemistry may have been the discovery of the first enzyme, diastase (today called amylase), in 1833 by Anselme Payen, while others considered Eduard Buchner's first demonstration of a complex biochemical process alcoholic fermentation in cell-free extracts in 1897 to be the birth of biochemistry. Some might also point as its beginning to the influential 1842 work by Justus von Liebig, Animal chemistry, or, Organic chemistry in its applications to physiology and pathology, which presented a chemical theory of metabolism, or even earlier to the 18th century studies on fermentation and respiration by Antoine Lavoisier. Many other pioneers in the field who helped to uncover the layers of complexity of biochemistry have been proclaimed founders of modern biochemistry, for example Emil Fischer for his work on the chemistry of proteins, and F. Gowland Hopkins on enzymes and the dynamic nature of biochemistry.

The term "biochemistry" itself is derived from a combination of biology and chemistry. In 1877, Felix Hoppe-Seyler used the term (biochemie in German) as a synonym for physiological chemistry in the foreword to the first issue of Zeitschrift für Physiologische Chemie (Journal of Physiological

Chemistry) where he argued for the setting up of institutes dedicated to this field of study. The German chemist Carl Neuberg however is often cited to have coined the word in 1903, while some credited it to Franz Hofmeister.

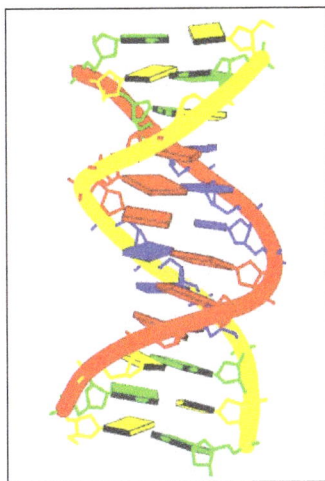

DNA structure (1D65).

It was once generally believed that life and its materials had some essential property or substance (often referred to as the "vital principle") distinct from any found in non-living matter, and it was thought that only living beings could produce the molecules of life. Then, in 1828, Friedrich Wöhler published a paper on the synthesis of urea, proving that organic compounds can be created artificially. Since then, biochemistry has advanced, especially since the mid-20th century, with the development of new techniques such as chromatography, X-ray diffraction, dual polarisation interferometry, NMR spectroscopy, radioisotopic labeling, electron microscopy, and molecular dynamics simulations. These techniques allowed for the discovery and detailed analysis of many molecules and metabolic pathways of the cell, such as glycolysis and the Krebs cycle (citric acid cycle), and led to an understanding of biochemistry on a molecular level. Philip Randle is well known for his discovery in diabetes research is possibly the glucose-fatty acid cycle in 1963. He confirmed that fatty acids reduce oxidation of sugar by the muscle. High fat oxidation was responsible for the insulin resistance.

Another significant historic event in biochemistry is the discovery of the gene, and its role in the transfer of information in the cell. This part of biochemistry is often called molecular biology of the gene. In the 1950s, James D. Watson, Francis Crick, Rosalind Franklin, and Maurice Wilkins were instrumental in solving DNA structure and suggesting its relationship with genetic transfer of information. In 1958, George Beadle and Edward Tatum received the Nobel Prize for work in fungi showing that one gene produces one enzyme. In 1988, Colin Pitchfork was the first person convicted of murder with DNA evidence, which led to the growth of forensic science. More recently, Andrew Z. Fire and Craig C. Mello received the 2006 Nobel Prize for discovering the role of RNA interference (RNAi), in the silencing of gene expression.

Starting Materials: The Chemical Elements of Life

Around two dozen of the 92 naturally occurring chemical elements are essential to various kinds of biological life. Most rare elements on Earth are not needed by life (exceptions being selenium and iodine), while a few common ones (aluminum and titanium) are not used. Most organisms share element needs, but there are a few differences between plants and animals. For example, ocean algae use bromine, but land plants and animals seem to need none. All animals require sodium, but some plants do not. Plants need boron and silicon, but animals may not (or may need ultra-small amounts).

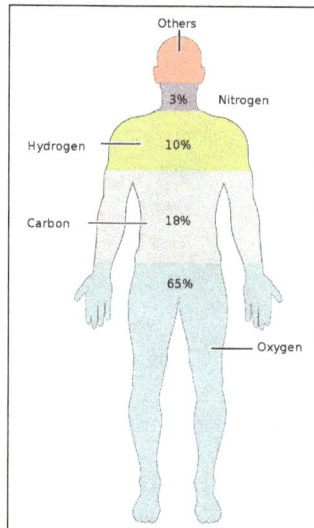

The main elements that compose the human body are shown
from most abundant (by mass) to least abundant.

Just six elements—carbon, hydrogen, nitrogen, oxygen, calcium, and phosphorus—make up almost 99% of the mass of living cells, including those in the human body. In addition to the six major elements that compose most of the human body, humans require smaller amounts of possibly 18 more.

Biomolecules

The four main classes of molecules in biochemistry (often called biomolecules) are carbohydrates, lipids, proteins, and nucleic acids. Many biological molecules are polymers: in this terminology, monomers are relatively small micromolecules that are linked together to create large macromolecules known as polymers. When monomers are linked together to synthesize a biological polymer, they undergo a process called dehydration synthesis. Different macromolecules can assemble in larger complexes, often needed for biological activity.

Carbohydrates

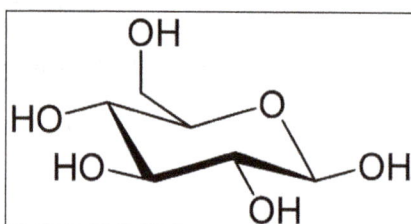

Glucose, a monosaccharide.

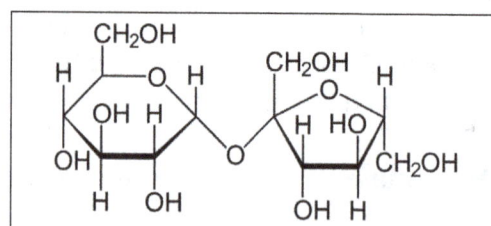

A molecule of sucrose (glucose + fructose),
a disaccharide.

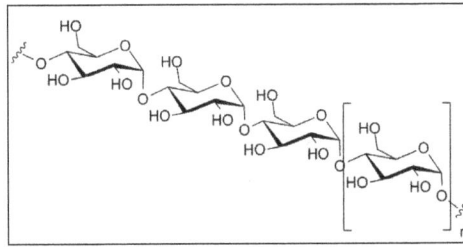

Amylose, a polysaccharide made up of
several thousand glucose units

Two of the main functions of carbohydrates are energy storage and providing structure. Sugars are carbohydrates, but not all carbohydrates are sugars. There are more carbohydrates on Earth than any other known type of biomolecule; they are used to store energy and genetic information, as well as play important roles in cell to cell interactions and communications.

The simplest type of carbohydrate is a monosaccharide, which among other properties contains carbon, hydrogen, and oxygen, mostly in a ratio of 1:2:1 (generalized formula $C_nH_{2n}O_n$, where n is at least 3). Glucose ($C_6H_{12}O_6$) is one of the most important carbohydrates; others include fructose ($C_6H_{12}O_6$), the sugar commonly associated with the sweet taste of fruits, and deoxyribose ($C_5H_{10}O_4$), a component of DNA. A monosaccharide can switch between acyclic (open-chain) form and a cyclic form. The open-chain form can be turned into a ring of carbon atoms bridged by an oxygen atom created from the carbonyl group of one end and the hydroxyl group of another. The cyclic molecule has a hemiacetal or hemiketal group, depending on whether the linear form was an aldose or a ketose.

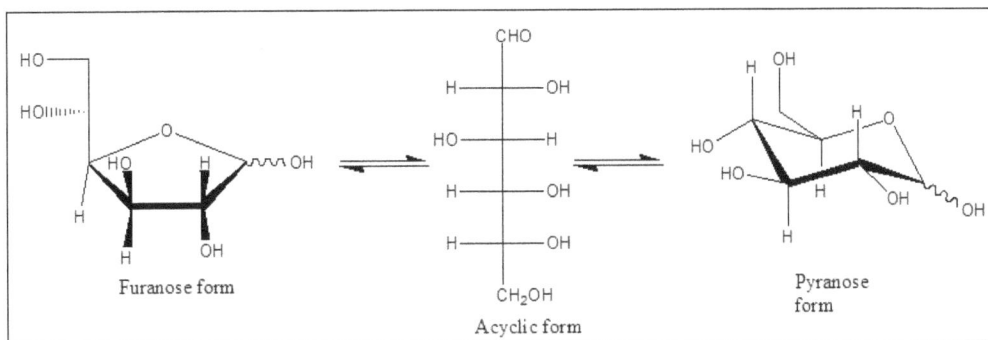

Conversion between the furanose, acyclic, and pyranose forms of D-glucose.

In these cyclic forms, the ring usually has 5 or 6 atoms. These forms are called furanoses and pyranoses, respectively—by analogy with furan and pyran, the simplest compounds with the same carbon-oxygen ring (although they lack the carbon-carbon double bonds of these two molecules). For example, the aldohexose glucose may form a hemiacetal linkage between the hydroxyl on carbon 1 and the oxygen on carbon 4, yielding a molecule with a 5-membered ring, called glucofuranose. The same reaction can take place between carbons 1 and 5 to form a molecule with a 6-membered ring, called glucopyranose. Cyclic forms with a 7-atom ring called heptoses are rare.

Two monosaccharides can be joined together by a glycosidic or ether bond into a *disaccharide* through a dehydration reaction during which a molecule of water is released. The reverse reaction in which the glycosidic bond of a disaccharide is broken into two monosaccharides is termed *hydrolysis*. The best-known disaccharide is sucrose or ordinary sugar, which consists of a glucose molecule and a fructose molecule joined together. Another important disaccharide is lactose found in milk, consisting of a glucose molecule and a galactose molecule. Lactose may be hydrolysed by lactase, and deficiency in this enzyme results in lactose intolerance.

When a few (around three to six) monosaccharides are joined, it is called an *oligosaccharide* (*oligo-* meaning "few"). These molecules tend to be used as markers and signals, as well as having some other uses. Many monosaccharides joined together make a polysaccharide. They can be joined together in one long linear chain, or they may be branched. Two of the most common polysaccharides are cellulose and glycogen, both consisting of repeating glucose monomers. *Cellulose* is an important structural component of plant's cell walls and *glycogen* is used as a form of energy storage in animals.

Sugar can be characterized by having reducing or non-reducing ends. A reducing end of a carbohydrate is a carbon atom that can be in equilibrium with the open-chain aldehyde (aldose) or keto form (ketose). If the joining of monomers takes place at such a carbon atom, the free hydroxy group of the pyranose or furanose form is exchanged with an OH-side-chain of another sugar, yielding a full acetal. This prevents opening of the chain to the aldehyde or keto form and renders the modified residue non-reducing. Lactose contains a reducing end at its glucose moiety, whereas the galactose moiety forms a full acetal with the C4-OH group of glucose. Saccharose does not have a reducing end because of full acetal formation between the aldehyde carbon of glucose (C1) and the keto carbon of fructose (C2).

Lipids

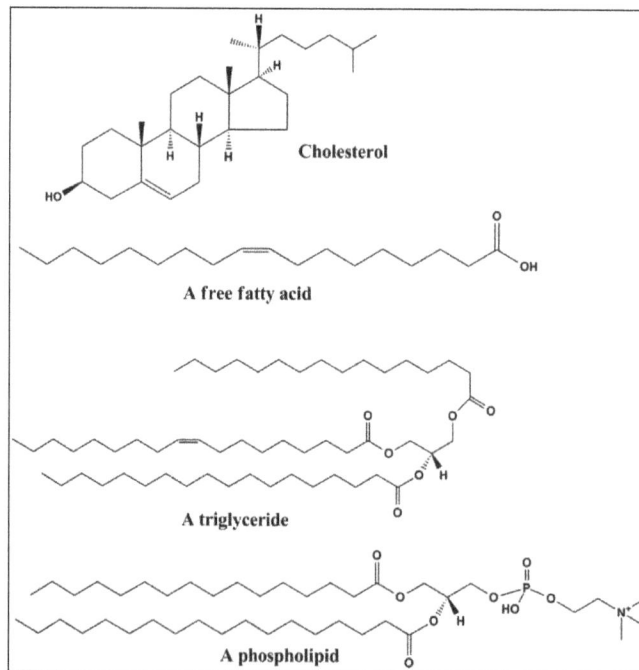

Structures of some common lipids.

At the top are cholesterol and oleic acid. The middle structure is a triglyceride composed of oleoyl, stearoyl, and palmitoyl chains attached to a glycerol backbone. At the bottom is the common phospholipid, phosphatidylcholine.

Lipids comprises a diverse range of molecules and to some extent is a catchall for relatively water-insoluble or nonpolar compounds of biological origin, including waxes, fatty acids, fatty-acid derived phospholipids, sphingolipids, glycolipids, and terpenoids (e.g., retinoids and steroids). Some lipids are linear, open chain aliphatic molecules, while others have ring structures. Some are aromatic (with a cyclic [ring] and planar [flat] structure) while others are not. Some are flexible, while others are rigid.

Lipids are usually made from one molecule of glycerol combined with other molecules. In triglycerides, the main group of bulk lipids, there is one molecule of glycerol and three fatty acids. Fatty acids are considered the monomer in that case, and may be saturated (no double bonds in the carbon chain) or unsaturated (one or more double bonds in the carbon chain).

Most lipids have some polar character in addition to being largely nonpolar. In general, the bulk of their structure is nonpolar or hydrophobic ("water-fearing"), meaning that it does not interact well with polar solvents like water. Another part of their structure is polar or hydrophilic ("water-loving") and will tend to associate with polar solvents like water. This makes them amphiphilic molecules (having both hydrophobic and hydrophilic portions). In the case of cholesterol, the polar group is a mere –OH (hydroxyl or alcohol). In the case of phospholipids, the polar groups are considerably larger and more polar, as described below.

Lipids are an integral part of our daily diet. Most oils and milk products that we use for cooking and eating like butter, cheese, ghee etc., are composed of fats. Vegetable oils are rich in various polyunsaturated fatty acids (PUFA). Lipid-containing foods undergo digestion within the body and are broken into fatty acids and glycerol, which are the final degradation products of fats and lipids. Lipids, especially phospholipids, are also used in various pharmaceutical products, either as co-solubilisers (e.g., in parenteral infusions) or else as drug carrier components (e.g., in a liposome or transfersome).

Proteins

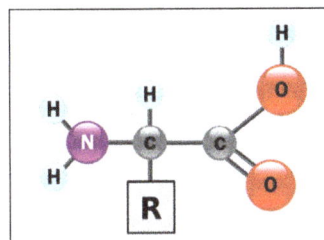

The general structure of an α-amino acid, with the amino group on the left and the carboxyl group on the right.

Proteins are very large molecules—macro-biopolymers—made from monomers called amino acids. An amino acid consists of a carbon atom attached to an amino group, $–NH_2$, a carboxylic acid group, $–COOH$ (although these exist as $–NH_3^+$ and $–COO^-$ under physiologic conditions), a simple hydrogen atom, and a side chain commonly denoted as "–R". The side chain "R" is different for each amino acid of which there are 20 standard ones. It is this "R" group that made each amino acid different, and the properties of the side-chains greatly influence the overall three-dimensional conformation of a protein. Some amino acids have functions by themselves or in a modified form; for instance, glutamate functions as an important neurotransmitter. Amino acids can be joined via a peptide bond. In this dehydration synthesis, a water molecule is removed and the peptide bond connects the nitrogen of one amino acid's amino group to the carbon of the other's carboxylic acid group. The resulting molecule is called a *dipeptide*, and short stretches of amino acids (usually, fewer than thirty) are called *peptides* or polypeptides. Longer stretches merit the title *proteins*. As an example, the important blood serum protein albumin contains 585 amino acid residues.

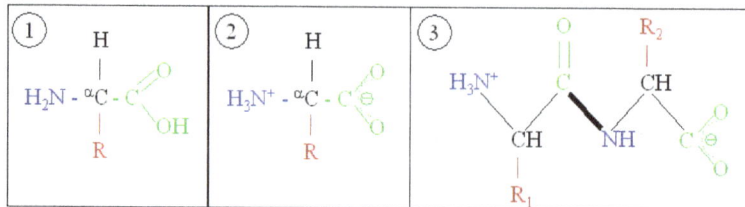

Generic amino acids (1) in neutral form, (2) as they exist physiologically, and (3) joined together as a dipeptide.

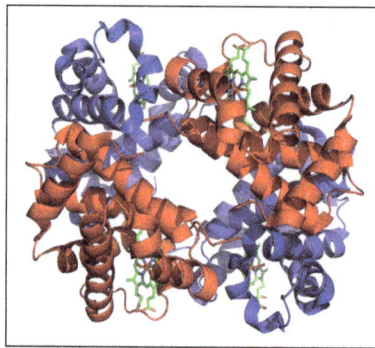

A schematic of hemoglobin. The red and blue ribbons represent the protein globin; the green structures are the heme groups.

Proteins can have structural and/or functional roles. For instance, movements of the proteins actin and myosin ultimately are responsible for the contraction of skeletal muscle. One property many proteins have is that they specifically bind to a certain molecule or class of molecules—they may be extremely selective in what they bind. Antibodies are an example of proteins that attach to one specific type of molecule. Antibodies are composed of heavy and light chains. Two heavy chains would be linked to two light chains through disulfide linkages between their amino acids. Antibodies are specific through variation based on differences in the N-terminal domain.

In fact, the enzyme-linked immunosorbent assay (ELISA), which uses antibodies, is one of the most sensitive tests modern medicine uses to detect various biomolecules. Probably the most important proteins, however, are the enzymes. Virtually every reaction in a living cell requires an enzyme to lower the activation energy of the reaction. These molecules recognize specific reactant molecules called substrates; they then catalyze the reaction between them. By lowering the activation energy, the enzyme speeds up that reaction by a rate of 10^{11} or more; a reaction that would normally take over 3,000 years to complete spontaneously might take less than a second with an enzyme. The enzyme itself is not used up in the process, and is free to catalyze the same reaction with a new set of substrates. Using various modifiers, the activity of the enzyme can be regulated, enabling control of the biochemistry of the cell as a whole.

The structure of proteins is traditionally described in a hierarchy of four levels. The primary structure of a protein consists of its linear sequence of amino acids; for instance, "alanine-glycine-tryptophan-serine-glutamate-asparagine-glycine-lysine-". Secondary structure is concerned with local morphology (morphology being the study of structure). Some combinations of amino acids will tend to curl up in a coil called an α-helix or into a sheet called a β-sheet; some α-helixes can be seen in the hemoglobin schematic above. Tertiary structure is the entire three-dimensional shape of the protein. This shape is determined by the sequence of amino acids. In fact, a single change can change the entire structure. The alpha chain of hemoglobin contains 146 amino acid residues; substitution of the glutamate residue at position 6 with a valine residue changes the behavior of hemoglobin so much that it results in sickle-cell disease. Finally, quaternary structure is concerned with the structure of a protein with multiple peptide subunits, like hemoglobin with its four subunits. Not all proteins have more than one subunit.

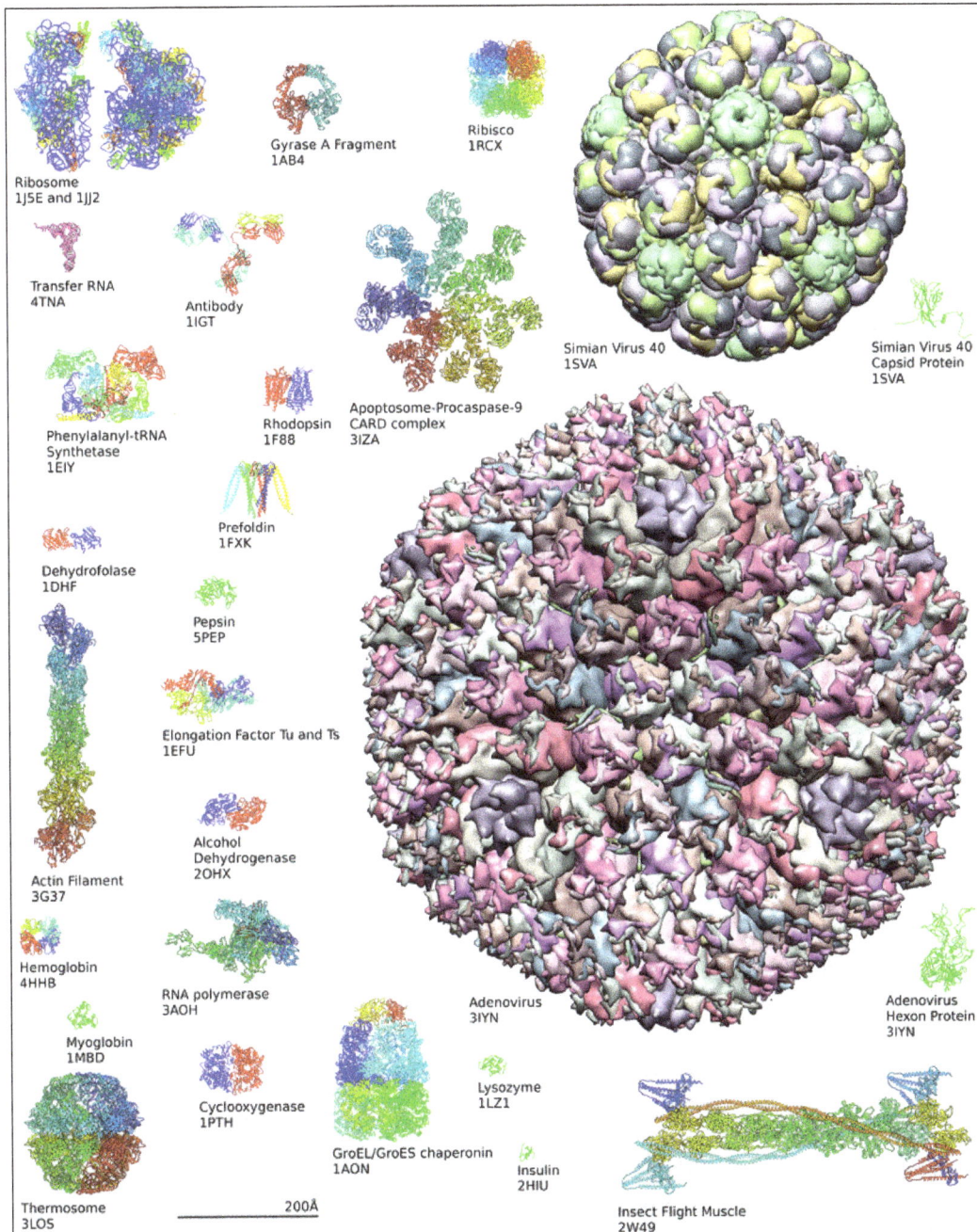

Examples of protein structures from the Protein Data Bank.

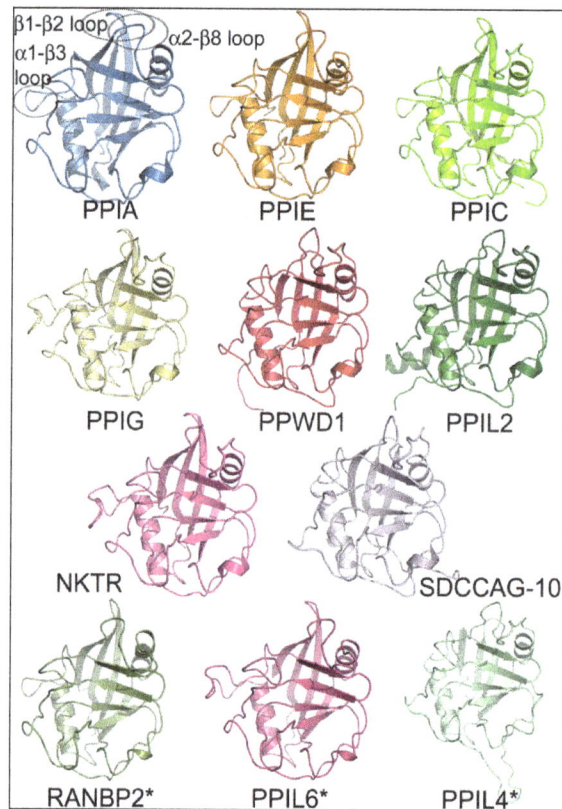

β1-β2 loop
α1-β3 loop
α2-β8 loop
PPIA   PPIE   PPIC
PPIG   PPWD1   PPIL2
NKTR   SDCCAG-10
RANBP2*   PPIL6*   PPIL4*

Members of a protein family, as represented by
the structures of the isomerase domains.

Ingested proteins are usually broken up into single amino acids or dipeptides in the small intestine, and then absorbed. They can then be joined to make new proteins. Intermediate products of glycolysis, the citric acid cycle, and the pentose phosphate pathway can be used to make all twenty amino acids, and most bacteria and plants possess all the necessary enzymes to synthesize them. Humans and other mammals, however, can synthesize only half of them. They cannot synthesize isoleucine, leucine, lysine, methionine, phenylalanine, threonine, tryptophan, and valine. These are the essential amino acids, since it is essential to ingest them. Mammals do possess the enzymes to synthesize alanine, asparagine, aspartate, cysteine, glutamate, glutamine, glycine, proline, serine, and tyrosine, the nonessential amino acids. While they can synthesize arginine and histidine, they cannot produce it in sufficient amounts for young, growing animals, and so these are often considered essential amino acids.

If the amino group is removed from an amino acid, it leaves behind a carbon skeleton called an α-keto acid. Enzymes called transaminases can easily transfer the amino group from one amino acid (making it an α-keto acid) to another α-keto acid (making it an amino acid). This is important in the biosynthesis of amino acids, as for many of the pathways, intermediates from other biochemical pathways are converted to the α-keto acid skeleton, and then an amino group is added, often via transamination. The amino acids may then be linked together to make a protein.

A similar process is used to break down proteins. It is first hydrolyzed into its component amino acids. Free ammonia (NH_3), existing as the ammonium ion (NH_4^+) in blood, is toxic to life forms. A suitable method for excreting it must therefore exist. Different tactics have evolved in different animals, depending on the animals' needs. Unicellular organisms simply release the ammonia into the environment. Likewise, bony fish can release the ammonia into the water where it is quickly diluted. In general, mammals convert the ammonia into urea, via the urea cycle.

In order to determine whether two proteins are related, or in other words to decide whether they are homologous or not, scientists use sequence-comparison methods. Methods like sequence alignments and structural alignments are powerful tools that help scientists identify homologies between related molecules. The relevance of finding homologies among proteins goes beyond forming an evolutionary pattern of protein families. By finding how similar two protein sequences are, we acquire knowledge about their structure and therefore their function.

Nucleic Acids

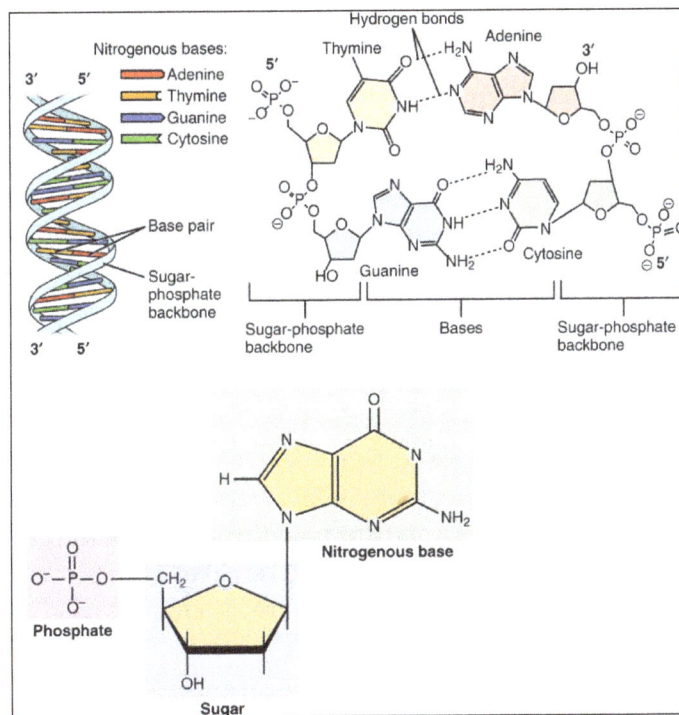

The structure of deoxyribonucleic acid (DNA),
the picture shows the monomers being put together.

Nucleic acids, so called because of their prevalence in cellular nuclei, is the generic name of the family of biopolymers. They are complex, high-molecular-weight biochemical macromolecules that can convey genetic information in all living cells and viruses. The monomers are called nucleotides, and each consists of three components: a nitrogenous heterocyclic base (either a purine or a pyrimidine), a pentose sugar, and a phosphate group.

The most common nucleic acids are deoxyribonucleic acid (DNA) and ribonucleic acid (RNA). The phosphate group and the sugar of each nucleotide bond with each other to form the backbone of the nucleic acid, while the sequence of nitrogenous bases stores the information. The most common nitrogenous bases are adenine, cytosine, guanine, thymine, and uracil. The nitrogenous bases of each strand of a nucleic acid will form hydrogen bonds with certain other nitrogenous bases in a complementary strand of nucleic acid (similar to a zipper). Adenine binds with thymine and uracil; thymine binds only with adenine; and cytosine and guanine can bind only with one another.

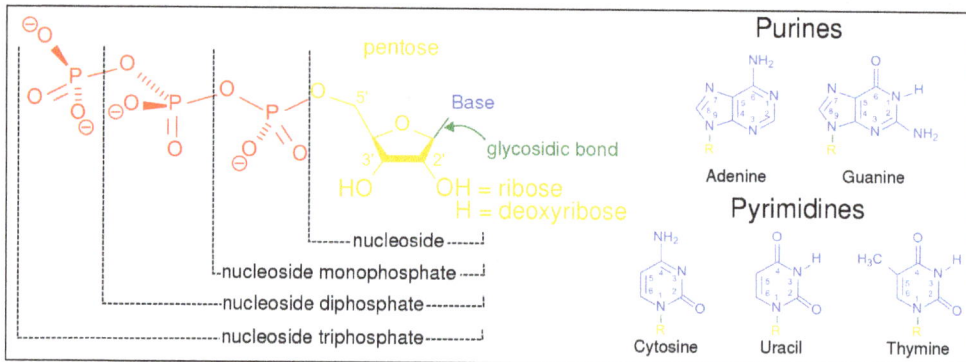

Structural elements of common nucleic acid constituents.

Because they contain at least one phosphate group, the compounds marked nucleoside mono-phosphate, nucleoside diphosphate and nucleoside triphosphate are all nucleotides (not simply phosphate-lacking nucleosides).

Aside from the genetic material of the cell, nucleic acids often play a role as second messengers, as well as forming the base molecule for adenosine triphosphate (ATP), the primary energy-carrier molecule found in all living organisms. Also, the nitrogenous bases possible in the two nucleic acids are different: adenine, cytosine, and guanine occur in both RNA and DNA, while thymine occurs only in DNA and uracil occurs in RNA.

Metabolism

Carbohydrates as Energy Source

Glucose is an energy source in most life forms. For instance, polysaccharides are broken down into their monomers by enzymes (glycogen phosphorylase removes glucose residues from glycogen, a polysaccharide). Disaccharides like lactose or sucrose are cleaved into their two component mono-saccharides.

Glycolysis (Anaerobic)

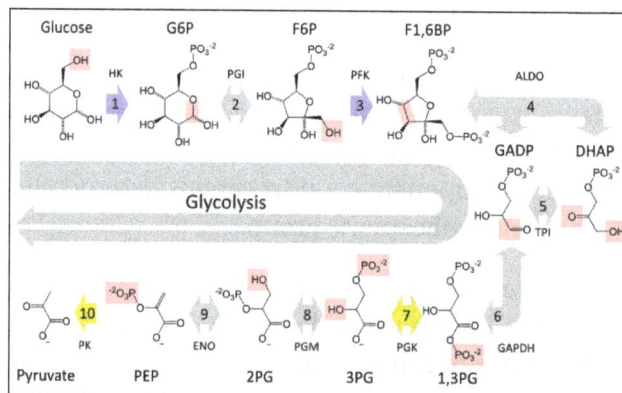

The metabolic pathway of glycolysis converts glucose to pyruvate by via a series of intermediate metabolites.

Each chemical modification (red box) is performed by a different enzyme. Steps 1 and 3 consume ATP (blue) and steps 7 and 10 produce ATP (yellow). Since steps 6-10 occur twice per glucose molecule, this leads to a net production of ATP.

Glucose is mainly metabolized by a very important ten-step pathway called glycolysis, the net result of which is to break down one molecule of glucose into two molecules of pyruvate. This also produces a net two molecules of ATP, the energy currency of cells, along with two reducing equivalents of converting NAD^+ (nicotinamide adenine dinucleotide: oxidised form) to NADH (nicotinamide adenine dinucleotide: reduced form). This does not require oxygen; if no oxygen is available (or the cell cannot use oxygen), the NAD is restored by converting the pyruvate to lactate (lactic acid) (e.g., in humans) or to ethanol plus carbon dioxide (e.g., in yeast). Other monosaccharides like galactose and fructose can be converted into intermediates of the glycolytic pathway.

Aerobic

In aerobic cells with sufficient oxygen, as in most human cells, the pyruvate is further metabolized. It is irreversibly converted to acetyl-CoA, giving off one carbon atom as the waste product carbon dioxide, generating another reducing equivalent as NADH. The two molecules acetyl-CoA (from one molecule of glucose) then enter the citric acid cycle, producing two molecules of ATP, six more NADH molecules and two reduced (ubi)quinones (via $FADH_2$ as enzyme-bound cofactor), and releasing the remaining carbon atoms as carbon dioxide. The produced NADH and quinol molecules then feed into the enzyme complexes of the respiratory chain, an electron transport system transferring the electrons ultimately to oxygen and conserving the released energy in the form of a proton gradient over a membrane (inner mitochondrial membrane in eukaryotes). Thus, oxygen is reduced to water and the original electron acceptors NAD^+ and quinone are regenerated. This is why humans breathe in oxygen and breathe out carbon dioxide. The energy released from transferring the electrons from high-energy states in NADH and quinol is conserved first as proton gradient and converted to ATP via ATP synthase. This generates an additional *28* molecules of ATP (24 from the 8 NADH + 4 from the 2 quinols), totaling to 32 molecules of ATP conserved per degraded glucose (two from glycolysis + two from the citrate cycle). It is clear that using oxygen to completely oxidize glucose provides an organism with far more energy than any oxygen-independent metabolic feature, and this is thought to be the reason why complex life appeared only after Earth's atmosphere accumulated large amounts of oxygen.

Gluconeogenesis

In vertebrates, vigorously contracting skeletal muscles (during weightlifting or sprinting, for example) do not receive enough oxygen to meet the energy demand, and so they shift to anaerobic metabolism, converting glucose to lactate. The liver regenerates the glucose, using a process called gluconeogenesis. This process is not quite the opposite of glycolysis, and actually requires three times the amount of energy gained from glycolysis (six molecules of ATP are used, compared to the two gained in glycolysis). Analogous to the above reactions, the glucose produced can then undergo glycolysis in tissues that need energy, be stored as glycogen (or starch in plants), or be converted to other monosaccharides or joined into di- or oligosaccharides. The combined pathways of glycolysis during exercise, lactate's crossing via the bloodstream to the liver, subsequent gluconeogenesis and release of glucose into the bloodstream is called the Cori cycle.

Relationship to other "Molecular-scale" Biological Sciences

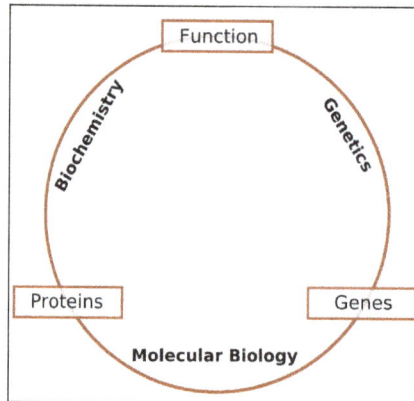

Schematic relationship between biochemistry, genetics, and molecular biology.

Researchers in biochemistry use specific techniques native to biochemistry, but increasingly combine these with techniques and ideas developed in the fields of genetics, molecular biology and biophysics. There is not a defined line between these disciplines. Biochemistry studies the chemistry required for biological activity of molecules, molecular biology studies their biological activity, genetics studies their heredity, which happens to be carried by their genome. This is shown in the following schematic that depicts one possible view of the relationships between the fields:

- Biochemistry is the study of the chemical substances and vital processes occurring in live organisms. Biochemists focus heavily on the chemistry behind the role, function, and structure of biomolecules. The study of the chemistry behind biological processes and the synthesis of biologically active molecules are examples of biochemistry. Biochemistry studies the chemistry which gives rise to the biology of molecules studied by molecular biology.

- Genetics is the study of the effect of genetic differences in organisms. This can often be inferred by the absence of a normal component (e.g. one gene). The study of "mutants" – organisms which lack one or more functional components with respect to the so-called "wild type" or normal phenotype. Genetic interactions (epistasis) can often confound simple interpretations of such "knockout" studies. Genetics studies the heredity to daughter cell or organism of the biology of macromolecules which is studied by molecular biology.

- Molecular biology is the study of molecular underpinnings of biological activity. It studies the structure, function, processing, regulation, interactions and evolution of biomolecules. Its most famous subfield, molecular genetics, studies the processes of replication, transcription, translation, and genetic mechanisms of cell function. The central dogma of molecular biology where genetic material is transcribed into RNA and then translated into protein, despite being oversimplified, still provides a good starting point for understanding the field. The picture has been revised in light of emerging novel roles for RNA.

- 'Chemical biology' seeks to develop new tools based on small molecules that allow minimal perturbation of biological systems while providing detailed information about their function. Further, chemical biology employs biological systems to create non-natural hybrids between biomolecules and synthetic devices (for example emptied viral capsids that can deliver gene therapy or drug molecules).

Biophysics

Biophysics is an interdisciplinary science that applies approaches and methods traditionally used in physics to study biological phenomena. Biophysics covers all scales of biological organization, from molecular to organismic and populations. Biophysical research shares significant overlap with biochemistry, molecular biology, physical chemistry, physiology, nanotechnology, bioengineering, computational biology, biomechanics, developmental biology and systems biology.

The term biophysics was originally introduced by Karl Pearson in 1892. Ambiguously, the term biophysics is also regularly used in academia to indicate the study of the physical quantities (e.g. electric current, temperature, stress, entropy) in biological systems, which is, by definition, performed by physiology. Nevertheless, other biological sciences also perform research on the biophysical properties of living organisms including molecular biology, cell biology, biophysics, and biochemistry.

Molecular biophysics typically addresses biological questions similar to those in biochemistry and molecular biology, seeking to find the physical underpinnings of biomolecular phenomena. Scientists in this field conduct research concerned with understanding the interactions between the various systems of a cell, including the interactions between DNA, RNA and protein biosynthesis, as well as how these interactions are regulated. A great variety of techniques are used to answer these questions.

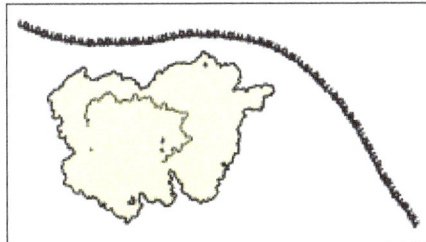

A ribosome is a biological machine that utilizes protein dynamics.

Fluorescent imaging techniques, as well as electron microscopy, x-ray crystallography, NMR spectroscopy, atomic force microscopy (AFM) and small-angle scattering (SAS) both with X-rays and neutrons (SAXS/SANS) are often used to visualize structures of biological significance. Protein dynamics can be observed by neutron spin echo spectroscopy. Conformational change in structure can be measured using techniques such as dual polarisation interferometry, circular dichroism, SAXS and SANS. Direct manipulation of molecules using optical tweezers or AFM, can also be used to monitor biological events where forces and distances are at the nanoscale. Molecular biophysicists often consider complex biological events as systems of interacting entities which can be understood e.g. through statistical mechanics, thermodynamics and chemical kinetics. By drawing knowledge and experimental techniques from a wide variety of disciplines, biophysicists are often able to directly observe, model or even manipulate the structures and interactions of individual molecules or complexes of molecules.

In addition to traditional (i.e. molecular and cellular) biophysical topics like structural biology or enzyme kinetics, modern biophysics encompasses an extraordinarily broad range of research, from bioelectronics to quantum biology involving both experimental and theoretical tools. It is becoming increasingly common for biophysicists to apply the models and experimental techniques derived from physics, as well as mathematics and statistics, to larger systems such as tissues, organs, populations and ecosystems. Biophysical models are used extensively in the study of electrical conduction in single neurons, as well as neural circuit analysis in both tissue and whole brain.

Medical physics, a branch of biophysics, is any application of physics to medicine or healthcare, ranging from radiology to microscopy and nanomedicine. For example, physicist Richard Feynman theorized about the future of nanomedicine. He wrote about the idea of a medical use for biological machines. Feynman and Albert Hibbs suggested that certain repair machines might one day be reduced in size to the point that it would be possible to (as Feynman put it) "swallow the doctor". The idea was discussed in Feynman's 1959 essay There's Plenty of Room at the Bottom.

Focus as a Subfield

While some colleges and universities have dedicated departments of biophysics, usually at the graduate level, many do not have university-level biophysics departments, instead having groups in related departments such as biochemistry, cell biology, chemistry, computer science, engineering, mathematics, medicine, molecular biology, neuroscience, pharmacology, physics, and physiology. Depending on the strengths of a department at a university differing emphasis will be given to fields of biophysics. What follows is a list of examples of how each department applies its efforts toward the study of biophysics. This list is hardly all inclusive. Nor does each subject of study belong exclusively to any particular department. Each academic institution makes its own rules and there is much overlap between departments.

- Biology and molecular biology – Gene regulation, single protein dynamics, bioenergetics, patch clamping, biomechanics, virophysics.

- Structural biology – Ångstrom-resolution structures of proteins, nucleic acids, lipids, carbohydrates, and complexes thereof.

- Biochemistry and chemistry – Biomolecular structure, siRNA, nucleic acid structure, structure-activity relationships.

- Computer science – Neural networks, biomolecular and drug databases.

- Computational chemistry – Molecular dynamics simulation, molecular docking, quantum chemistry.

- Bioinformatics – Sequence alignment, structural alignment, protein structure prediction.

- Mathematics – Graph/network theory, population modeling, dynamical systems, phylogenetics.

- Medicine – Biophysical research that emphasizes medicine. Medical biophysics is a field closely related to physiology. It explains various aspects and systems of the body from a physical and mathematical perspective. Examples are fluid dynamics of blood flow, gas physics of respiration, radiation in diagnostics/treatment and much more. Biophysics is taught as a preclinical subject in many medical schools, mainly in Europe.

- Neuroscience – Studying neural networks experimentally (brain slicing) as well as theoretically (computer models), membrane permittivity, gene therapy, understanding tumors.

- Pharmacology and physiology – Channelomics, biomolecular interactions, cellular membranes, polyketides.

- Physics – Negentropy, stochastic processes, and the development of new physical techniques and instrumentation as well as their application.

- Quantum biology – The field of quantum biology applies quantum mechanics to biological objects and problems. Decohered isomers to yield time-dependent base substitutions. These studies imply applications in quantum computing.

- Agronomy and agriculture.

Many biophysical techniques are unique to this field. Research efforts in biophysics are often initiated by scientists who were biologists, chemists or physicists by training.

Developemental Biology

Developmental biology is the study of the process by which animals and plants grow and develop. Developmental biology also encompasses the biology of regeneration, asexual reproduction, metamorphosis, and the growth and differentiation of stem cells in the adult organism.

In the late 20th century, the discipline largely transformed into evolutionary developmental biology.

Perspectives

The main processes involved in the embryonic development of animals are: regional specification, morphogenesis, cell differentiation, growth, and the overall control of timing explored in evolutionary developmental biology:

- Regional specification refers to the processes that create spatial pattern in a ball or sheet of initially similar cells. This generally involves the action of cytoplasmic determinants, located within parts of the fertilized egg, and of inductive signals emitted from signaling centers in the embryo. The early stages of regional specification do not generate functional differentiated cells, but cell populations committed to develop to a specific region or part of the organism. These are defined by the expression of specific combinations of transcription factors.

- Morphogenesis relates to the formation of three-dimensional shape. It mainly involves the orchestrated movements of cell sheets and of individual cells. Morphogenesis is important for creating the three germ layers of the early embryo (ectoderm, mesoderm and endoderm) and for building up complex structures during organ development.

- Cell differentiation relates specifically to the formation of functional cell types such as nerve, muscle, secretory epithelia etc. Differentiated cells contain large amounts of specific proteins associated with the cell function.

- Growth involves both an overall increase in size, and also the differential growth of parts (allometry) which contributes to morphogenesis. Growth mostly occurs through cell division but also through changes of cell size and the deposition of extracellular materials.

- The control of timing of events and the integration of the various processes with one another is the least well understood area of the subject. It remains unclear whether animal embryos contain a master clock mechanism or not.

The development of plants involves similar processes to that of animals. However plant cells are mostly immotile so morphogenesis is achieved by differential growth, without cell movements. Also, the inductive signals and the genes involved are different from those that control animal development.

Developmental Processes

Cell Differentiation

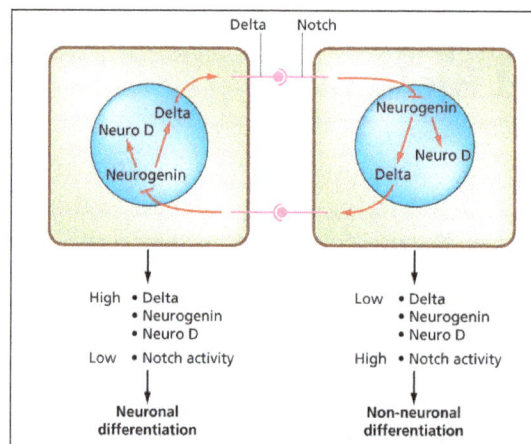

The Notch-delta system in neurogenesis.

Cell differentiation is the process whereby different functional cell types arise in development. For example, neurons, muscle fibers and hepatocytes (liver cells) are well known types of differentiated cells. Differentiated cells usually produce large amounts of a few proteins that are required for their specific function and this gives them the characteristic appearance that enables them to be recognized under the light microscope. The genes encoding these proteins are highly active. Typically their chromatin structure is very open, allowing access for the transcription enzymes, and specific transcription factors bind to regulatory sequences in the DNA in order to activate gene expression. For example, NeuroD is a key transcription factor for neuronal differentiation, myogenin for muscle differentiation, and HNF4 for hepatocyte differentiation.

Cell differentiation is usually the final stage of development, preceded by several states of commitment which are not visibly differentiated. A single tissue, formed from a single type of progenitor cell or stem cell, often consists of several differentiated cell types. Control of their formation involves a process of lateral inhibition, based on the properties of the Notch signaling pathway. For example, in the neural plate of the embryo this system operates to generate a population of neuronal precursor cells in which NeuroD is highly expressed.

Regeneration

Regeneration indicates the ability to regrow a missing part. This is very prevalent amongst plants, which show continuous growth, and also among colonial animals such as hydroids and ascidians. But most interest by developmental biologists has been shown in the regeneration of parts in free living animals. In particular four models have been the subject of much investigation. Two of these have the ability to regenerate whole bodies: *Hydra*, which can regenerate any part of the polyp from a small fragment, and planarian worms, which can usually regenerate both heads and tails. Both of these examples have continuous cell turnover fed by stem cells and, at least in planaria, at least some of the stem cells have been shown to be pluripotent. The other two models show only distal regeneration of appendages. These are the insect appendages, usually the legs of hemimetabolous insects such as the cricket, and the limbs of urodele amphibians. Considerable information is now available about amphibian limb regeneration and it is known that each cell type regenerates itself, except for connective tissues where there is considerable interconversion between cartilage, dermis and tendons. In terms of the pattern of structures, this is controlled by a re-activation of signals active in the embryo. There is still debate about the old question of whether regeneration is a "pristine" or an "adaptive" property. If the former is the case, with improved knowledge, we might expect to be able to improve regenerative ability in humans. If the latter, then each instance of regeneration is presumed to have arisen by natural selection in circumstances particular to the species, so no general rules would be expected.

Embryonic Development of Animals

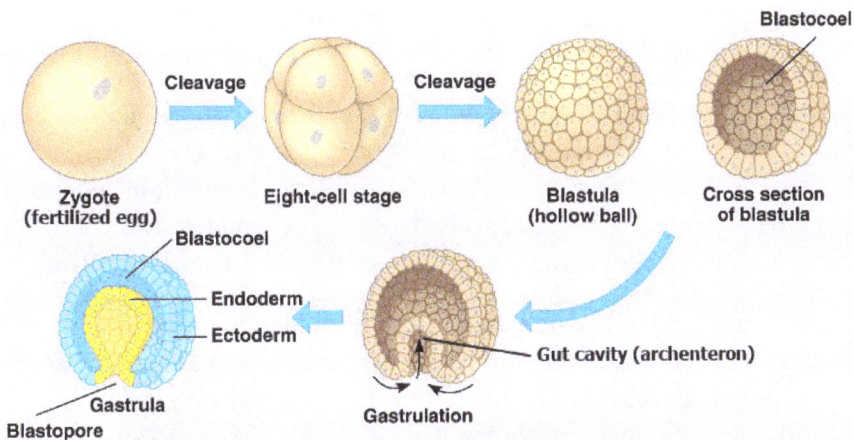

Generalized scheme of embryonic development.

The sperm and egg fuse in the process of fertilization to form a fertilized egg, or zygote. This undergoes a period of divisions to form a ball or sheet of similar cells called a blastula or blastoderm. These cell divisions are usually rapid with no growth so the daughter cells are half the size of the mother cell and the whole embryo stays about the same size. They are called cleavage divisions.

Mouse epiblast primordial germ cells undergo extensive epigenetic reprogramming. This process involves genome-wide DNA demethylation, chromatin reorganization and epigenetic imprint erasure leading to totipotency. DNA demethylation is carried out by a process that utilizes the DNA base excision repair pathway.

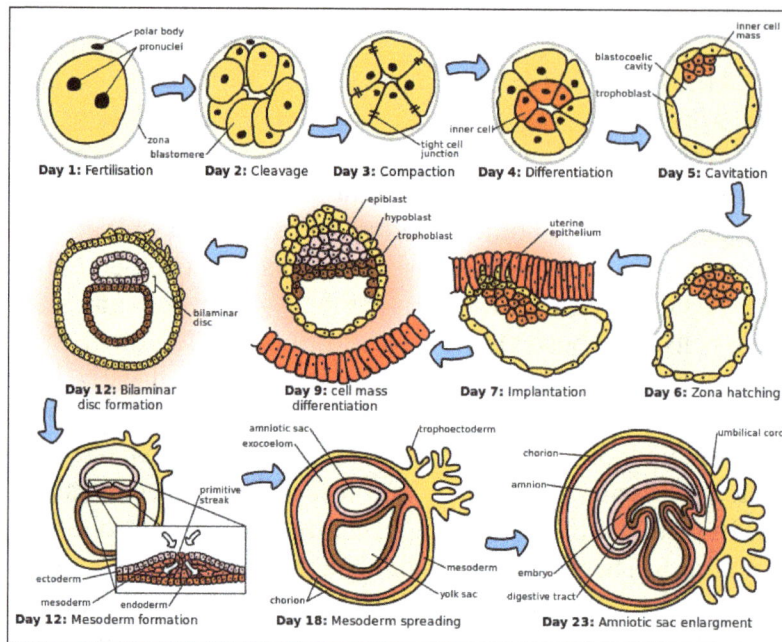

The initial stages of human embryogenesis.

Mouse epiblast primordial germ cells undergo extensive epigenetic reprogramming. This process involves genome-wide DNA demethylation, chromatin reorganization and epigenetic imprint erasure leading to totipotency. DNA demethylation is carried out by a process that utilizes the DNA base excision repair pathway.

Morphogenetic movements convert the cell mass into a three layered structure consisting of multicellular sheets called ectoderm, mesoderm and endoderm, which are known as germ layers. This is the process of gastrulation. During cleavage and gastrulation the first regional specification events occur. In addition to the formation of the three germ layers themselves, these often generate extraembryonic structures, such as the mammalian placenta, needed for support and nutrition of the embryo, and also establish differences of commitment along the anteroposterior axis (head, trunk and tail).

Regional specification is initiated by the presence of cytoplasmic determinants in one part of the zygote. The cells that contain the determinant become a signaling center and emit an inducing factor. Because the inducing factor is produced in one place, diffuses away, and decays, it forms a concentration gradient, high near the source cells and low further away. The remaining cells of the embryo, which do not contain the determinant, are competent to respond to different concentrations by upregulating specific developmental control genes. This results in a series of zones becoming set up, arranged at progressively greater distance from the signaling center. In each zone a different combination of developmental control genes is upregulated. These genes encode transcription factors which upregulate new combinations of gene activity in each region. Among other functions, these transcription factors control expression of genes conferring specific adhesive and motility properties on the cells in which they are active. Because of these different morphogenetic properties, the cells of each germ layer move to form sheets such that the ectoderm ends up on the outside, mesoderm in the middle, and endoderm on the inside. Morphogenetic movements not only change the shape and structure of the embryo, but by bringing cell sheets into new spatial relationships they also make possible new phases of signaling and response between them.

Growth in embryos is mostly autonomous. For each territory of cells the growth rate is controlled by the combination of genes that are active. Free-living embryos do not grow in mass as they have no external food supply. But embryos fed by a placenta or extraembryonic yolk supply can grow very fast, and changes to relative growth rate between parts in these organisms help to produce the final overall anatomy.

The whole process needs to be coordinated in time and how this is controlled is not understood. There may be a master clock able to communicate with all parts of the embryo that controls the course of events, or timing may depend simply on local causal sequences of events.

Metamorphosis

Developmental processes are very evident during the process of metamorphosis. This occurs in various types of animal. Well-known are the examples of the frog, which usually hatches as a tadpole and metamorphoses to an adult frog, and certain insects which hatch as a larva and then become remodeled to the adult form during a pupal stage.

All the developmental processes listed above occur during metamorphosis. Examples that have been especially well studied include tail loss and other changes in the tadpole of the frog Xenopus, and the biology of the imaginal discs, which generate the adult body parts of the fly Drosophila melanogaster.

Plant Development

Plant development is the process by which structures originate and mature as a plant grows. It is studied in plant anatomy and plant physiology as well as plant morphology.

Plants constantly produce new tissues and structures throughout their life from meristems located at the tips of organs, or between mature tissues. Thus, a living plant always has embryonic tissues. By contrast, an animal embryo will very early produce all of the body parts that it will ever have in its life. When the animal is born (or hatches from its egg), it has all its body parts and from that point will only grow larger and more mature.

The properties of organization seen in a plant are emergent properties which are more than the sum of the individual parts. "The assembly of these tissues and functions into an integrated multicellular organism yields not only the characteristics of the separate parts and processes but also quite a new set of characteristics which would not have been predictable on the basis of examination of the separate parts."

Growth

A vascular plant begins from a single celled zygote, formed by fertilisation of an egg cell by a sperm cell. From that point, it begins to divide to form a plant embryo through the process of embryogenesis. As this happens, the resulting cells will organize so that one end becomes the first root, while the other end forms the tip of the shoot. In seed plants, the embryo will develop one or more "seed leaves" (cotyledons). By the end of embryogenesis, the young plant will have all the parts necessary to begin in its life.

Once the embryo germinates from its seed or parent plant, it begins to produce additional organs (leaves, stems, and roots) through the process of organogenesis. New roots grow from root meristems located at the tip of the root, and new stems and leaves grow from shoot meristems located at the tip of the shoot. Branching occurs when small clumps of cells left behind by the meristem, and which have not yet undergone cellular differentiation to form a specialized tissue, begin to grow as the tip of a new root or shoot. Growth from any such meristem at the tip of a root or shoot is termed primary growth and results in the lengthening of that root or shoot. Secondary growth results in widening of a root or shoot from divisions of cells in a cambium.

In addition to growth by cell division, a plant may grow through cell elongation. This occurs when individual cells or groups of cells grow longer. Not all plant cells will grow to the same length. When cells on one side of a stem grow longer and faster than cells on the other side, the stem will bend to the side of the slower growing cells as a result. This directional growth can occur via a plant's response to a particular stimulus, such as light (phototropism), gravity (gravitropism), water, (hydrotropism), and physical contact (thigmotropism).

Plant growth and development are mediated by specific plant hormones and plant growth regulators (PGRs). Endogenous hormone levels are influenced by plant age, cold hardiness, dormancy, and other metabolic conditions; photoperiod, drought, temperature, and other external environmental conditions; and exogenous sources of PGRs, e.g., externally applied and of rhizospheric origin.

Morphological Variation

Plants exhibit natural variation in their form and structure. While all organisms vary from individual to individual, plants exhibit an additional type of variation. Within a single individual, parts are repeated which may differ in form and structure from other similar parts. This variation is most easily seen in the leaves of a plant, though other organs such as stems and flowers may show similar variation. There are three primary causes of this variation: positional effects, environmental effects, and juvenility.

Evolution of Plant Morphology

Transcription factors and transcriptional regulatory networks play key roles in plant morphogenesis and their evolution. During plant landing, many novel transcription factor families emerged and are preferentially wired into the networks of multicellular development, reproduction, and organ development, contributing to more complex morphogenesis of land plants.

Developmental Model Organisms

Much of developmental biology research in recent decades has focused on the use of a small number of model organisms. It has turned out that there is much conservation of developmental mechanisms across the animal kingdom. In early development different vertebrate species all use essentially the same inductive signals and the same genes encoding regional identity. Even invertebrates use a similar repertoire of signals and genes although the body parts formed are significantly different. Model organisms each have some particular experimental advantages which have enabled them to become popular among researchers. In one sense they are "models" for the whole animal kingdom, and in another sense they are "models" for human development, which is difficult to study directly for both ethical and practical reasons. Model organisms have been most useful for elucidating the broad nature of developmental mechanisms. The more detail is sought, the more they differ from each other and from humans.

Plants:

- Thale cress (*Arabidopsis thaliana*).

Vertebrates:

- Frog: Xenopus (X.laevis and tropicalis). Good embryo supply. Especially suitable for microsurgery.

- Zebrafish: Danio rerio. Good embryo supply. Well developed genetics.

- Chicken: Gallus gallus. Early stages similar to mammal, but microsurgery easier. Low cost.

- Mouse: Mus musculus. A mammal with well developed genetics.

Invertebrates:

- Fruit fly: Drosophila melanogaster. Good embryo supply. Well developed genetics.

- Nematode: Caenorhabditis elegans. Good embryo supply. Well developed genetics. Low cost.

Also popular for some purposes have been sea urchins and ascidians. For studies of regeneration urodele amphibians such as the axolotl *Ambystoma mexicanum* are used, and also planarian worms such as *Schmidtea mediterranea*. Organoids have also been demonstrated as an efficient model for development. Plant development has focused on the thale cress *Arabidopsis thaliana* as a model organism.

Embrology

Embryology is the branch of biology that studies the prenatal development of gametes (sex cells), fertilization, and development of embryos and fetuses. Additionally, embryology encompasses the study of congenital disorders that occur before birth, known as teratology.

Embryology has a long history. Aristotle proposed the currently accepted theory of epigenesis, that organisms develop from seed or egg in a sequence of steps. The alternative theory, preformationism, that organisms develop from pre-existing miniature versions of themselves, however, held sway until the 18th century. Modern embryology developed from the work of von Baer, though accurate observations had been made in Italy by anatomists such as Aldrovandi and Leonardo da Vinci in the Renaissance.

Embryonic Development of Animals

After cleavage, the dividing cells, or morula, becomes a hollow ball, or blastula, which develops a hole or pore at one end.

Bilateria

In bilateral animals, the blastula develops in one of two ways that divide the whole animal kingdom into two halves. If in the blastula the first pore (blastopore) becomes the mouth of the animal, it is a protostome; if the first pore becomes the anus then it is a deuterostome. The protostomes include most invertebrate animals, such as insects, worms and molluscs, while the deuterostomes include the vertebrates. In due course, the blastula changes into a more differentiated structure called the gastrula.

The gastrula with its blastopore soon develops three distinct layers of cells (the germ layers) from which all the bodily organs and tissues then develop:

- The innermost layer, or endoderm, give rise to the digestive organs, the gills, lungs or swim bladder if present, and kidneys or nephrites.

- The middle layer, or mesoderm, gives rise to the muscles, skeleton if any, and blood system.

- The outer layer of cells, or ectoderm, gives rise to the nervous system, including the brain, and skin or carapace and hair, bristles, or scales.

Embryos in many species often appear similar to one another in early developmental stages. The reason for this similarity is because species have a shared evolutionary history. These similarities among species are called homologous structures, which are structures that have the same or similar function and mechanism, having evolved from a common ancestor.

Drosophila Melanogaster (Fruit Fly)

Drosophila melanogaster.

Drosophila melanogaster, a fruit fly, is a model organism in biology on which much research into embryology has been done. Before fertilization, the female gamete produces an abundance of mRNA - transcribed from the genes that encode bicoid protein and nanos protein. These mRNA molecules are stored to be used later in what will become the developing embryo. The male and female Drosophila gametes exhibit anisogamy (differences in morphology and sub-cellular biochemistry). The female gamete is larger than the male gamete because it harbors more cytoplasm and, within the cytoplasm, the female gamete contains an abundance of the mRNA previously mentioned. At fertilization, the male and female gametes fuse (plasmogamy) and then the nucleus of the male gamete fuses with the nucleus of the female gamete (karyogamy). Note that before the gametes' nuclei fuse, they are known as pronuclei. A series of nuclear divisions will occur without cytokinesis (division of the cell) in the zygote to form a multi-nucleated cell (a cell containing multiple nuclei) known as a syncytium. All the nuclei in the syncytium are identical, just as all the nuclei in every somatic cell of any multicellular organism are identical in terms of the DNA sequence of the genome. Before the nuclei can differentiate in transcriptional activity, the embryo (syncytium) must be divided into segments. In each segment, a unique set of regulatory proteins will cause specific genes in the nuclei to be transcribed. The resulting combination of proteins will transform clusters of cells into early embryo tissues that will each develop into multiple fetal and adult tissues later in development (note: this happens after each nucleus becomes wrapped with its own cell membrane).

Drosophila melanogaster larvae contained in lab apparatus
to be used for experiments in genetics and embryology.

Outlined below is the process that leads to cell and tissue differentiation.

Maternal-effect genes - subject to Maternal (cytoplasmic) inheritance.

- Egg-polarity genes establish the Anteroposterior axis.

Zygotic-effect genes - subject to Mendelian (classical) inheritance

- Segmentation genes establish 14 segments of the embryo using the anteroposterior axis as a guide.

  ◦ Gap genes establish 3 broad segments of the embryo.

  ◦ Pair-rule genes define 7 segments of the embryo within the confines of the second broad segment that was defined by the gap genes.

  ◦ Segment-polarity genes define another 7 segments by dividing each of the pre-existing 7 segments into anterior and posterior halves.

- Homeotic (homeobox) genes use the 14 segments as pinpoints for specific types of cell differentiation and the histological developments that correspond to each cell type.

Humans

Humans are bilaterals and deuterostomes. In humans, the term embryo refers to the ball of dividing cells from the moment the zygote implants itself in the uterus wall until the end of the eighth week after conception. Beyond the eighth week after conception (tenth week of pregnancy), the developing human is then called a fetus.

Vertebrate and Invertebrate Embryology

Many principles of embryology apply to invertebrates as well as to vertebrates. Therefore, the study of invertebrate embryology has advanced the study of vertebrate embryology. However, there are many differences as well. For example, numerous invertebrate species release a larva before development is complete; at the end of the larval period, an animal for the first time comes to resemble an adult similar to its parent or parents. Although invertebrate embryology is similar in some ways for different invertebrate animals, there are also countless variations. For instance, while spiders proceed directly from egg to adult form, many insects develop through at least one larval stage.

Gerontology

Gerontology is the study of the social, cultural, psychological, cognitive, and biological aspects of ageing. The word was coined by Ilya Ilyich Mechnikov in 1903. The field is distinguished from geriatrics, which is the branch of medicine that specializes in the treatment of existing disease in older adults. Gerontologists include researchers and practitioners in the fields of biology, nursing, medicine, criminology, dentistry, social work, physical and occupational therapy, psychology, psychiatry, sociology, economics, political science, architecture, geography, pharmacy, public health, housing, and anthropology.

The multidisciplinary nature of gerontology means that there are a number of sub-fields which overlap with gerontology. There are policy issues, for example, involved in government planning and the operation of nursing homes, investigating the effects of an ageing population on society, and the design of residential spaces for older people that facilitate the development of a sense of place or home. Dr. Lawton, a behavioral psychologist at the Philadelphia Geriatric Center, was among the first to recognize the need for living spaces designed to accommodate the elderly, especially those with Alzheimer's disease. As an academic discipline the field is relatively new. The USC Leonard Davis School created the first PhD, master's and bachelor's degree programs in gerontology in 1975.

Aging Demographics

The world is forecast to undergo rapid population aging in the next several decades. In 1900, there were 3.1 million people aged 65 years and older living in the United States. However, this population continued to grow throughout the 20th century and reached 31.2, 35, and 40.3 million people in 1990, 2000, and 2010, respectively. Notably, in the United States and across the world, the "baby boomer" generation began to turn 65 in 2011. Recently, the population aged 65 years and older has grown at a faster rate than the total population in the United States. The total population increased by 9.7%, from 281.4 million to 308.7 million, between 2000 and 2010. However, the population aged 65 years and older increased by 15.1% during the same period. It has been estimated that 25% of the population in the United States and Canada will be aged 65 years and older by 2025. Moreover, by 2050, it is predicted that, for the first time in United States history, the number of individuals aged 60 years and older will be greater than the number of children aged 0 to 14 years. Those aged 85 years and older (oldest-old) are projected to increase from 5.3 million to 21 million by 2050. Adults aged 85–89 years constituted the greatest segment of the oldest-old in 1990, 2000, and 2010. However, the largest percentage point increase among the oldest-old occurred in the 90- to 94-year-old age group, which increased from 25.0% in 1990 to 26.4% in 2010.

With the rapid growth of the aging population, social work education and training specialized in older adults and practitioners interested in working with older adults are increasingly in demand.

Gender Differences with Age

There has been a considerable disparity between the number of men and women in the older population in the United States. In both 2000 and 2010, women outnumbered men in the older population at every single year of age (e.g., 65 to 100 years and over). The sex ratio, which is a measure used to indicate the balance of males to females in a population, is calculated by taking the number of males divided by the number of females, and multiplying by 100. Therefore, the sex ratio is the number of males per 100 females. In 2010, there were 90.5 males per 100 females in the 65-year-old population. However, this represented an increase from 1990 when there were 82.7 males per 100 females, and from 2000 when the sex ratio was 88.1. Although the gender gap between men and women has narrowed, women continue to have a greater life expectancy and lower mortality rates at older ages relative to men. For example, the Census 2010 reported that there were approximately twice as many women as men living in the United States at 89 years of age (361,309 versus 176,689, respectively).

Geographic Distribution of Older Adults

The number and percentage of older adults living in the United States vary across the four different regions (Northeast, Midwest, West, and South) defined by the United States census. In 2010, the South contained the greatest number of people aged 65 years and older and 85 years and older. However, proportionately, the Northeast contains the largest percentage of adults aged 65 years and older (14.1%), followed by the Midwest (13.5%), the South (13.0%), and the West (11.9%). Relative to the Census 2000, all geographic regions demonstrated positive growth in the population of adults aged 65 years and older and 85 years and older. The most rapid growth in the population of adults aged 65 years and older was evident in the West (23.5%), which showed an increase from 6.9 million in 2000 to 8.5 million in 2010. Likewise, in the population aged 85 years and older, the West (42.8%) also showed the fastest growth and increased from 806,000 in 2000 to 1.2 million in 2010. It is worth highlighting that Rhode Island was the only state that experienced a reduction in the number of people aged 65 years and older, and declined from 152,402 in 2000 to 151,881 in 2010. Conversely, all states exhibited an increase in the population of adults aged 85 years and older from 2000 to 2010.

Biogerontology

Biogerontology is the sub-field of gerontology concerned with the biological aging process, its evolutionary origins, and potential means to intervene in the process. It involves interdisciplinary research on biological aging's causes, effects, and mechanisms. Conservative biogerontologists such as Leonard Hayflick have predicted that the human life expectancy will peak at about 92 years old, while others such as James Vaupel have predicted that in industrialized countries, life expectancies will reach 100 for children born after the year 2000. and some surveyed biogerontologists have predicted life expectancies of two or more centuries. with Aubrey de Grey offering the "tentative timeframe" that with adequate funding of research to develop interventions in aging such as Strategies for Engineered Negligible Senescence, "we have a 50/50 chance of developing technology within about 25 to 30 years from now that will, under reasonable assumptions about the rate of subsequent improvements in that technology, allow us to stop people from dying of aging at any age", leading to life expectancies of 1,000 years.

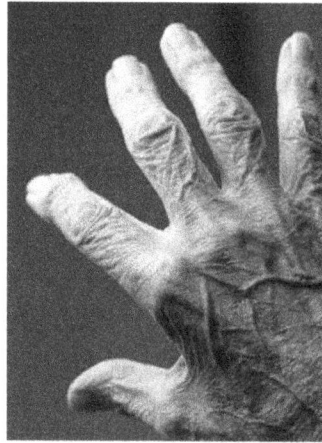

The hand of an older adult.

Biomedical gerontology, also known as experimental gerontology and life extension, is a sub-discipline of biogerontology that endeavors to slow, prevent, and even reverse aging in both humans and animals. Most "life extensionists" believe the human life span can be increased within the next century, if not sooner. Biogerontologists vary in the degree to which they focus on the study of the aging process as a means of mitigating the diseases of aging or extending lifespan, although most agree that extension of lifespan will necessarily flow from reductions in age-related disease and frailty, although some argue that maximum life span cannot be altered or that it is undesirable to try. Geroscience is a recently formulated interdisciplinary field that embraces biomedical gerontology as the center of preventing diseases of aging through science.

In contrast with biogerontology, which aims to prevent age-related disease by intervening in aging processes, geriatrics is a field of medicine that studies the treatment of existing disease in aging people.

There are numerous theories of aging, and no one theory has been accepted. There is a wide spectrum of the types of theories for the causes of aging with programmed theories on one extreme and error theories on the other. Regardless of the theory, a commonality is that as humans age, functions of the body decline.

Stochastic Theories

Stochastic theories of aging is the suggestion that aging is caused by small changes in the body over time and the body's failure to restore the system and mend the damages to the body. The cells and tissues are eventually injured due to the damage gathered over time. This causes the diminishes in an organ's function related to age. The notion of accumulated damage was first introduced by Weisman as the "wear and tear" theory.

Wear and Tear Theory

Wear and tear theories of aging suggest that as an individual ages, body parts such as cells and organs wear out from continued use. Wearing of the body can be attributable to internal or external causes that eventually lead to an accumulation of insults which surpasses the capacity for repair. Due to these internal and external insults, cells lose their ability to regenerate, which ultimately leads to mechanical and chemical exhaustion. Some insults include chemicals in the air, food, or smoke. Other insults may be things such as viruses, trauma, free radicals, cross-linking, and high body temperature.

Accumulation

Accumulation theories of aging suggest that aging is bodily decline that results from an accumulation of elements, whether introduced to the body from the environment or resulting from cell metabolism. An example of an accumulation theory is the Free Radical Theory of Aging.

Free Radical Theory

Free radicals are reactive molecules produced by cellular and environmental processes, and can damage the elements of the cell such as the cell membrane and DNA and cause irreversible damage. The free-radical theory of aging proposes that this damage cumulatively degrades the biological function of cells and impacts the process of aging. The idea that free radicals are toxic agents was first proposed by Rebeca Gerschman and colleagues in 1945, but came to prominence in 1956, when Denham Harman proposed the free-radical theory of aging and even demonstrated that free radical reactions contribute to the degradation of biological systems. Oxidative damage of many types accumulate with age, such as oxidative stress that oxygen-free radicals, because the free radical theory of aging argues that aging results from the damage generated by reactive oxygen species (ROS). ROS are small, highly reactive, oxygen-containing molecules that can damage a complex of cellular components such as fat, proteins, or from DNA, they are naturally generated in small amounts during the body's metabolic reactions. These conditions become more common as humans grow older and include diseases related to aging, such as dementia, cancer and heart disease.

DNA Damage Theory

DNA damage has been one of the many causes in diseases related to aging. The stability of the genome is defined by the cells machinery of repair, damage tolerance, and checkpoint pathways that counteracts DNA damage. One hypothesis proposed by Gioacchino Failla in 1958 is that damage accumulation to the DNA causes aging. The hypothesis was developed soon by physicist Leó Szilárd. This theory has changed over the years as new research has discovered new types of DNA damage and mutations, and several theories of aging argue that DNA damage with or without mutations causes aging.

Cross-linking Theory

The cross-linking theory proposes that advanced glycation end-products (stable bonds formed by the binding of glucose to proteins) and other aberrant cross-links accumulating in aging tissues is the cause of aging. The crosslinking of proteins disables their biological functions. The hardening of the connective tissue, kidney diseases, and enlargement of the heart are connected to the cross-linking of proteins. Crosslinking of DNA can induce replication errors, and this leads to deformed cells and increases the risk of cancer.

Genetic

Genetic theories of aging propose that aging is programmed within each individual's genes. According to this theory, genes dictate cellular longevity. Programmed cell death, or apoptosis, is determined by a "biological clock" via genetic information in the nucleus of the cell. Genes responsible for apoptosis provide an explanation for cell death, but are less applicable to death of an entire organism. An increase in cellular apoptosis may correlate to aging, but is not a 'cause of death'. Environmental factors and genetic mutations can influence gene expression and accelerate aging. More recently epigenetics have been explored as a contributing factor. The epigenetic clock, which objectively measures the biological age of cells and tissues, may become useful for testing different biological aging theories.

General Imbalance

General imbalance theories of aging suggest that body systems, such as the endocrine, nervous, and immune systems, gradually decline and ultimately fail to function. The rate of failure varies system by system.

Immunological Theory

The immunological theory of aging suggests that the immune system weakens as an organism ages. This makes the organism unable to fight infections and less able to destroy old and neoplastic cells. This leads to aging and will eventually lead to death. This theory of aging was developed by Ray Walford, an American gerontologist. According to Walford, incorrect immunological procedures are the cause of the process of aging.

Social Gerontology

Social gerontology is a multi-disciplinary sub-field that specializes in studying or working with older adults. Social gerontologists may have degrees or training in social work, nursing, psychology, sociology, demography, public health, or other social science disciplines. Social gerontologists are responsible for educating, researching, and advancing the broader causes of older people.

Because issues of life span and life extension need numbers to quantify them, there is an overlap with demography. Those who study the demography of the human life span differ from those who study the social demographics of aging.

Social Theories of Aging

Several theories of aging are developed to observe the aging process of older adults in society as well as how these processes are interpreted by men and women as they age.

Activity Theory

Activity theory was developed and elaborated by Cavan, Havighurst, and Albrecht. According to this theory, older adults' self-concept depends on social interactions. In order for older adults to maintain morale in old age, substitutions must be made for lost roles. Examples of lost roles include retirement from a job or loss of a spouse.

Activity is preferable to inactivity because it facilitates well-being on multiple levels. Because of improved general health and prosperity in the older population, remaining active is more feasible now than when this theory was first proposed by Havighurst nearly six decades ago. The activity theory is applicable for a stable, post-industrial society, which offers its older members many opportunities for meaningful participation.Weakness: Some aging persons cannot maintain a middle-aged lifestyle, due to functional limitations, lack of income, or lack of a desire to do so. Many older adults lack the resources to maintain active roles in society. On the flip side, some elders may insist on continuing activities in late life that pose a danger to themselves and others, such as driving at night with low visual acuity or doing maintenance work to the house while climbing with severely arthritic knees. In doing so, they are denying their limitations and engaging in unsafe behaviors.

Disengagement Theory

Disengagement theory was developed by Cumming and Henry. According to this theory, older adults and society engage in a mutual separation from each other. An example of mutual separation is retirement from the workforce. A key assumption of this theory is that older adults lose "ego-energy" and become increasingly self-absorbed. Additionally, disengagement leads to higher morale maintenance than if older adults try to maintain social involvement. This theory is heavily criticized for having an escape clause - namely, that older adults who remain engaged in society are unsuccessful adjusters to old age.

Gradual withdrawal from society and relationships preserves social equilibrium and promotes self-reflection for elders who are freed from societal roles. It furnishes an orderly means for the transfer of knowledge, capital, and power from the older generation to the young. It makes it possible for society to continue functioning after valuable older members die.

Continuity Theory

Continuity theory is an elusive concept. On one hand, to exhibit continuity can mean to remain the same, to be uniform, homogeneous, unchanging, even humdrum. This static view of continuity is not very applicable to human aging. On the other hand, a dynamic view of continuity starts with the idea of a basic structure which persists over time, but it allows for a variety of changes to occur within the context provided by the basic structure. The basic structure is coherent: It has an orderly or logical relation of parts that is recognizably unique and that allows us to differentiate that structure from others. With the introduction of the concept of time, ideas such as direction, sequence, character development, and story line enter into the concept of continuity as it is applied to the evolution of a human being. In this theory, a dynamic concept of continuity is developed and applied to the issue of adaptation to normal aging.

A central premise of continuity theory is that, in making adaptive choices, middle-aged and older adults attempt to preserve and maintain existing internal and external structures and that they prefer to accomplish this objective by using continuity (i.e., applying familiar strategies in familiar arenas of life). In middle and later life, adults are drawn by the weight of past experience to use continuity as a primary adaptive strategy for dealing with changes associated with normal aging. To the extent that change builds upon, and has links to, the person's past, change is a part of continuity. As a result of both their own perceptions and pressures from the social environment, individuals who are adapting to normal aging are both predisposed and motivated toward inner psychological continuity as well as outward continuity of social behavior and circumstances.

Continuity theory views both internal and external continuity as robust adaptive strategies that are supported by both individual preference and social sanctions. Continuity theory consists of general adaptive principles that people who are normally aging could be expected to follow, explanations of how these principles work, and a specification of general areas of life in which these principles could be expected to apply. Accordingly, continuity theory has enormous potential as a general theory of adaptation to individual aging.

Age Stratification Theory

According to this theory, older adults born during different time periods form cohorts that define "age strata". There are two differences among strata: chronological age and historical experience. This theory makes two arguments. 1. Age is a mechanism for regulating behavior and as a result determines access to positions of power. 2. Birth cohorts play an influential role in the process of social change.

Life Course Theory

According to this theory, which stems from the Life Course Perspective, aging occurs from birth to death. Aging involves social, psychological, and biological processes. Additionally, aging experiences are shaped by cohort and period effects.

Also reflecting the life course focus, consider the implications for how societies might function when age-based norms vanish—a consequence of the deinstitutionalization of the life course— and suggest that these implications pose new challenges for theorizing aging and the life course in postindustrial societies. Dramatic reductions in mortality, morbidity, and fertility over the past several decades have so shaken up the organization of the life course and the nature of educational, work, family, and leisure experiences that it is now possible for individuals to become old in new ways. The configurations and content of other life stages are being altered as well, especially for women. In consequence, theories of age and aging will need to be reconceptualized.

Cumulative Advantage/Disadvantage Theory

According to this theory, which was developed beginning in the 1960s by Derek Price and Robert Merton and elaborated on by several researchers such as Dale Dannefer, inequalities have a tendency to become more pronounced throughout the aging process. A paradigm of this theory can be expressed in the adage "the rich get richer and the poor get poorer". Advantages and disadvantages in early life stages have a profound effect throughout the life span. However, advantages and disadvantages in middle adulthood have a direct influence on economic and health status in later life.

Environmental Gerontology

Environmental gerontology is a specialization within gerontology that seeks an understanding and interventions to optimize the relationship between aging persons and their physical and social environments.

The field emerged in the 1930s during the first studies on behavioral and social gerontology. In the 1970s and 1980s, research confirmed the importance of the physical and social environment in understanding the aging population and improved the quality of life in old age. Studies of environmental gerontology indicate that older people prefer to age in their immediate environment, whereas spatial experience and place attachment are important for understanding the process.

Some research indicates that the physical-social environment is related to the longevity and quality of life of the elderly. Precisely, the natural environment (such as natural therapeutic landscapes, therapeutic garden) contributes to active and healthy aging in the place.

Jurisprudential Gerontology

Jurisprudential gerontology (sometimes referred to as "geriatric jurisprudence") is a specialization within gerontology that looks into the ways laws and legal structures interact with the aging experience. The field started from legal scholars in the field of elder law, which found that looking into legal issues of older persons without a broader inter-disciplinary perspective does not provide the ideal legal outcome. Using theories such as therapeutic jurisprudence, jurisprudential scholars critically examined existing legal institutions (e.g. adult guardianship, end of life care, or nursing homes regulations) and showed how law should look more closely to the social and psychological aspects of its real-life operation. Other streams within jurisprudential gerontology also encouraged physicians and lawyers to try to improve their cooperation and better understand how laws and regulatory institutions affect health and well being of older persons.

Mathematical and Theoretical Biology

Mathematical and theoretical biology is a branch of biology which employs theoretical analysis, mathematical models and abstractions of the living organisms to investigate the principles that govern the structure, development and behavior of the systems, as opposed to experimental biology which deals with the conduction of experiments to prove and validate the scientific theories. The field is sometimes called mathematical biology or biomathematics to stress the mathematical side, or theoretical biology to stress the biological side. Theoretical biology focuses more on the development of theoretical principles for biology while mathematical biology focuses on the use of mathematical tools to study biological systems, even though the two terms are sometimes interchanged.

Mathematical biology aims at the mathematical representation and modeling of biological processes, using techniques and tools of applied mathematics and it can be useful in both theoretical and practical research. Describing systems in a quantitative manner means their behavior can be better simulated, and hence properties can be predicted that might not be evident to the experimenter. This requires precise mathematical models.

Because of the complexity of the living systems, theoretical biology employs several fields of mathematics, and has contributed to the development of new techniques.

Areas of Research

Several areas of specialized research in mathematical and theoretical biology as well as external links to related projects in various universities are concisely presented in the following subsections, including also a large number of appropriate validating references from a list of several thousands of published authors contributing to this field. Many of the included examples are characterised by highly complex, nonlinear, and supercomplex mechanisms, as it is being increasingly recognised that the result of such interactions may only be understood through a combination of mathematical, logical, physical/chemical, molecular and computational models.

Abstract Relational Biology

Abstract relational biology (ARB) is concerned with the study of general, relational models of complex biological systems, usually abstracting out specific morphological, or anatomical, structures. Some of the simplest models in ARB are the Metabolic-Replication, or (M,R)--systems introduced by Robert Rosen in 1957-1958 as abstract, relational models of cellular and organismal organization.

Other approaches include the notion of autopoiesis developed by Maturana and Varela, Kauffman's Work-Constraints cycles, and more recently the notion of closure of constraints.

Algebraic Biology

Algebraic biology (also known as symbolic systems biology) applies the algebraic methods of symbolic computation to the study of biological problems, especially in genomics, proteomics, analysis of molecular structures and study of genes.

Complex Systems Biology

An elaboration of systems biology to understanding the more complex life processes was developed since 1970 in connection with molecular set theory, relational biology and algebraic biology.

Computer Models and Automata Theory

A monograph on this topic summarizes an extensive amount of published research in this area up to 1986, including subsections in the following areas: computer modeling in biology and medicine, arterial system models, neuron models, biochemical and oscillation networks, quantum automata, quantum computers in molecular biology and genetics, cancer modelling, neural nets, genetic networks, abstract categories in relational biology, metabolic-replication systems, category theory applications in biology and medicine, automata theory, cellular automata, tessellation models and complete self-reproduction, chaotic systems in organisms, relational biology and organismic theories.

Modeling Cell and Molecular Biology

This area has received a boost due to the growing importance of molecular biology:

- Mechanics of biological tissues.

- Theoretical enzymology and enzyme kinetics.

- Cancer modelling and simulation.

- Modelling the movement of interacting cell populations.

- Mathematical modelling of scar tissue formation.

- Mathematical modelling of intracellular dynamics.

- Mathematical modelling of the cell cycle.

Modelling Physiological Systems

- Modelling of arterial disease.

- Multi-scale modelling of the heart.

- Modelling electrical properties of muscle interactions, as in bidomain and monodomain models.

Computational Neuroscience

Computational neuroscience (also known as theoretical neuroscience or mathematical neuroscience) is the theoretical study of the nervous system.

Evolutionary Biology

Ecology and evolutionary biology have traditionally been the dominant fields of mathematical biology.

Evolutionary biology has been the subject of extensive mathematical theorizing. The traditional approach in this area, which includes complications from genetics, is population genetics. Most population geneticists consider the appearance of new alleles by mutation, the appearance of new genotypes by recombination, and changes in the frequencies of existing alleles and genotypes at a small number of gene loci. When infinitesimal effects at a large number of gene loci are considered, together with the assumption of linkage equilibrium or quasi-linkage equilibrium, one derives quantitative genetics. Ronald Fisher made fundamental advances in statistics, such as analysis of variance, via his work on quantitative genetics. Another important branch of population genetics that led to the extensive development of coalescent theory is phylogenetics. Phylogenetics is an area that deals with the reconstruction and analysis of phylogenetic (evolutionary) trees and networks based on inherited characteristics Traditional population genetic models deal with alleles and genotypes, and are frequently stochastic.

Many population genetics models assume that population sizes are constant. Variable population sizes, often in the absence of genetic variation, are treated by the field of population dynamics. Work in this area dates back to the 19th century, and even as far as 1798 when Thomas Malthus formulated the first principle of population dynamics, which later became known as the Malthusian growth model. The Lotka–Volterra predator-prey equations are another famous example. Population dynamics overlap with another active area of research in mathematical biology: mathematical epidemiology, the study of infectious disease affecting populations. Various models of the spread of infections have been proposed and analyzed, and provide important results that may be applied to health policy decisions.

In evolutionary game theory, developed first by John Maynard Smith and George R. Price, selection acts directly on inherited phenotypes, without genetic complications. This approach has been mathematically refined to produce the field of adaptive dynamics.

Mathematical Biophysics

The earlier stages of mathematical biology were dominated by mathematical biophysics, described as the application of mathematics in biophysics, often involving specific physical/mathematical models of biosystems and their components or compartments.

The following is a list of mathematical descriptions and their assumptions.

Deterministic Processes (Dynamical Systems)

A fixed mapping between an initial state and a final state. Starting from an initial condition and moving forward in time, a deterministic process always generates the same trajectory, and no two trajectories cross in state space.

- Difference equations/Maps: Discrete time, continuous state space.
- Ordinary differential equations: Continuous time, continuous state space, no spatial derivatives.
- Partial differential equations: Continuous time, continuous state space, spatial derivatives.
- Logical deterministic cellular automata: Discrete time, discrete state space.

Stochastic Processes (Random Dynamical Systems)

A random mapping between an initial state and a final state, making the state of the system a random variable with a corresponding probability distribution.

- Non-Markovian processes: Generalized master equation – continuous time with memory of past events, discrete state space, waiting times of events (or transitions between states) discretely occur.
- Jump Markov process: Master equation – continuous time with no memory of past events, discrete state space, waiting times between events discretely occur and are exponentially distributed.
- Continuous Markov process: Stochastic differential equations or a Fokker-Planck equation – continuous time, continuous state space, events occur continuously according to a random Wiener process.

Spatial Modelling

One classic work in this area is Alan Turing's paper on morphogenesis entitled *The Chemical Basis of Morphogenesis*, published in 1952 in the Philosophical Transactions of the Royal Society.

- Travelling waves in a wound-healing assay.
- Swarming behavior.

- A mechanochemical theory of morphogenesis.

- Biological pattern formation.

- Spatial distribution modeling using plot samples.

- Turing patterns.

Mathematical Methods

A model of a biological system is converted into a system of equations, although the word 'model' is often used synonymously with the system of corresponding equations. The solution of the equations, by either analytical or numerical means, describes how the biological system behaves either over time or at equilibrium. There are many different types of equations and the type of behavior that can occur is dependent on both the model and the equations used. The model often makes assumptions about the system. The equations may also make assumptions about the nature of what may occur.

Molecular Set Theory

Molecular set theory (MST) is a mathematical formulation of the wide-sense chemical kinetics of biomolecular reactions in terms of sets of molecules and their chemical transformations represented by set-theoretical mappings between molecular sets. It was introduced by Anthony Bartholomay, and its applications were developed in mathematical biology and especially in mathematical medicine. In a more general sense, MST is the theory of molecular categories defined as categories of molecular sets and their chemical transformations represented as set-theoretical mappings of molecular sets. The theory has also contributed to biostatistics and the formulation of clinical biochemistry problems in mathematical formulations of pathological, biochemical changes of interest to Physiology, Clinical Biochemistry and Medicine.

Organizational Biology

Theoretical approaches to biological organization aim to understand the interdependence between the parts of organisms. They emphasize the circularities that these interdependences lead to. Theoretical biologists developed several concepts to formalize this idea.

For example, abstract relational biology (ARB) is concerned with the study of general, relational models of complex biological systems, usually abstracting out specific morphological, or anatomical, structures. Some of the simplest models in ARB are the Metabolic-Replication, or (M,R)--systems introduced by Robert Rosen in 1957-1958 as abstract, relational models of cellular and organismal organization.

Plant Biology

Botany is the branch of biology that deals with the study of plants, including their structure, properties, and biochemical processes. Also included are plant classification and the study of plant diseases and of interactions with the environment. The principles and findings of botany have provided the base for such applied sciences as agriculture, horticulture, and forestry.

Plants were of paramount importance to early humans, who depended upon them as sources of food, shelter, clothing, medicine, ornament, tools, and magic. Today it is known that, in addition to their practical and economic values, green plants are indispensable to all life on Earth: through the process of photosynthesis, plants transform energy from the Sun into the chemical energy of food, which makes all life possible. A second unique and important capacity of green plants is the formation and release of oxygen as a by-product of photosynthesis. The oxygen of the atmosphere, so absolutely essential to many forms of life, represents the accumulation of over 3,500,000,000 years of photosynthesis by green plants and algae.

Although the many steps in the process of photosynthesis have become fully understood only in recent years, even in prehistoric times humans somehow recognized intuitively that some important relation existed between the Sun and plants. Such recognition is suggested by the fact that worship of the Sun was often combined with the worship of plants by early tribes and civilizations.

Earliest humans, like the other anthropoid mammals (e.g., apes, monkeys), depended totally upon the natural resources of the environment, which, until methods were developed for hunting, consisted almost completely of plants. The behavior of pre-Stone Age humans can be inferred by studying the botany of aboriginal peoples in various parts of the world. Isolated tribal groups in South America, Africa, and New Guinea, for example, have extensive knowledge about plants and distinguish hundreds of kinds according to their utility, as edible, poisonous, or otherwise important in their culture. They have developed sophisticated systems of nomenclature and classification, which approximate the binomial system (i.e., generic and specific names) found in modern biology. The urge to recognize different kinds of plants and to give them names thus seems to be as old as the human race.

In time plants were not only collected but also grown by humans. This domestication resulted not only in the development of agriculture but also in a greater stability of human populations that had previously been nomadic. From the settling down of agricultural peoples in places where they could depend upon adequate food supplies came the first villages and the earliest civilizations.

Because of the long preoccupation of humans with plants, a large body of folklore, general information, and actual scientific data has accumulated, which has become the basis for the science of botany.

Areas of Study

For convenience, but not on any mutually exclusive basis, several major areas or approaches are recognized commonly as disciplines of botany. These are morphology, physiology, ecology, and systematics.

Morphology

Morphology deals with the structure and form of plants and includes such subdivisions as: cytology, the study of the cell; histology, the study of tissues; anatomy, the study of the organization of tissues into the organs of the plant; reproductive morphology, the study of life cycles; and experimental morphology, or morphogenesis, the study of development.

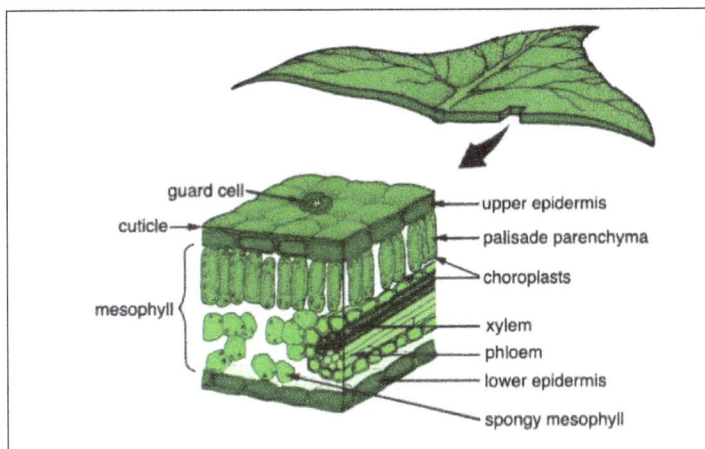

Structures of a leaf.

The epidermis is often covered with a waxy protective cuticle that helps prevent water loss from inside the leaf. Oxygen, carbon dioxide, and water enter and exit the leaf through pores (stomata) scattered mostly along the lower epidermis. The stomata are opened and closed by the contraction and expansion of surrounding guard cells. The vascular, or conducting, tissues are known as xylem and phloem; water and minerals travel up to the leaves from the roots through the xylem, and sugars made by photosynthesis are transported to other parts of the plant through the phloem. Photosynthesis occurs within the chloroplast-containing mesophyll layer.

Physiology

Physiology deals with the functions of plants. Its development as a subdiscipline has been closely interwoven with the development of other aspects of botany, especially morphology. In fact, structure and function are sometimes so closely related that it is impossible to consider one independently of the other. The study of function is indispensable for the interpretation of the incredibly diverse nature of plant structures. In other words, around the functions of the plant, structure and form have evolved. Physiology also blends imperceptibly into the fields of biochemistry and biophysics, as the research methods of these fields are used to solve problems in plant physiology.

Ecology

Ecology deals with the mutual relationships and interactions between organisms and their physical environment. The physical factors of the atmosphere, the climate, and the soil affect the physiological functions of the plant in all its manifestations, so that, to a large degree, plant ecology is a phase of plant physiology under natural and uncontrolled conditions. Plants are intensely sensitive to the forces of the environment, and both their association into communities and their geographical distribution are determined largely by the character of climate and soil. Moreover, the pressures of the environment and of organisms upon each other are potent forces, which lead to new species and the continuing evolution of larger groups. Ecology also investigates the competitive or mutualistic relationships that occur at different levels of ecosystem composition, such as those between individuals, populations, or communities. Plant-animal interactions, such as those between plants and their herbivores or pollinators, are also an important area of study.

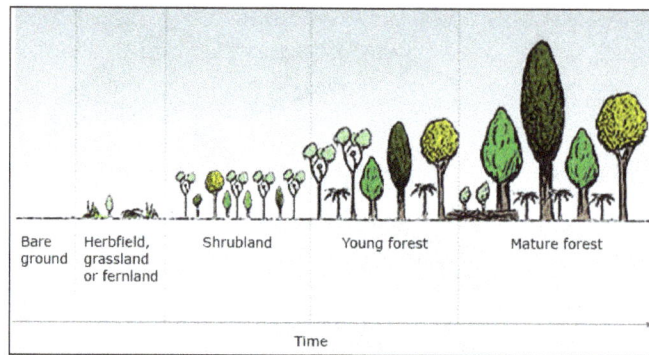

Primary succession.

Primary succession begins in barren areas, such as on bare rock exposed by a retreating glacier. The first inhabitants are lichens or plants—those that can survive in such an environment. Over hundreds of years these "pioneer species" convert the rock into soil that can support simple plants such as grasses. These grasses further modify the soil, which is then colonized by other types of plants. Each successive stage modifies the habitat by altering the amount of shade and the composition of the soil. The final stage of succession is a climax community, which is a very stable stage that can endure for hundreds of years.

Systematics

Systematics deals with the identification and ranking of all plants. It includes classification and nomenclature (naming) and enables the botanist to comprehend the broad range of plant diversity and evolution.

Other Subdisciplines

In addition to the major subdisciplines, several specialized branches of botany have developed as a matter of custom or convenience. Among them are bacteriology, the study of bacteria; mycology, the study of fungi; phycology, the study of algae; bryology, the study of mosses and liverworts; pteridology, the study of ferns and their relatives; and paleobotany, the study of fossil plants. Palynology is the study of modern and fossil pollen and spores, with particular reference to their identification; plant pathology deals with the diseases of plants; economic botany deals with plants of practical use to humankind; and ethnobotany covers the traditional use of plants by local peoples, now and in the distant past.

Botany also relates to other scientific disciplines in many ways, especially to zoology, medicine, microbiology, agriculture, chemistry, forestry, and horticulture, and specialized areas of botanical information may relate closely to such humanistic fields as art, literature, history, religion, archaeology, sociology, and psychology.

Fundamentally, botany remains a pure science, including any research into the life of plants and limited only by humanity's technical means of satisfying curiosity. It has often been considered an important part of a liberal education, not only because it is necessary for an understanding of agriculture, horticulture, forestry, pharmacology, and other applied arts and sciences but also because an understanding of plant life is related to life in general.

Because humanity has always been dependent upon plants and surrounded by them, plants are woven into designs, into the ornamentation of life, even into religious symbolism. A Persian carpet and a bedspread from a New England loom both employ conventional designs derived from the forms of flowers. Medieval painters and great masters of the Renaissance represented various revered figures surrounded by roses, lilies, violets, and other flowers, which symbolized chastity, martyrdom, humility, and other Christian attributes.

Methods in Botany

Morphological Aspects

The invention of the compound microscope provided a valuable and durable instrument for the investigation of the inner structure of plants. Early plant morphologists, especially those studying cell structure, were handicapped as much by the lack of adequate knowledge of how to prepare specimens as they were by the imperfect microscopes of the time. A revolution in the effectiveness of microscopy occurred in the second half of the 19th century with the introduction of techniques for fixing cells and for staining their component parts. Before the development of these techniques, the cell, viewed with the microscope, appeared as a minute container with a dense portion called the nucleus. The discovery that parts of the cell respond to certain stains made observation easier. The development of techniques for preparing tissues of plants for microscopic examination was continued in the 1870s and 1880s and resulted in the gradual refinement of the field of nuclear cytology, or karyology. Chromosomes were recognized as constant structures in the life cycle of cells, and the nature and meaning of meiosis, a type of cell division in which the daughter cells have half the number of chromosomes of the parent, was discovered; without this discovery, the significance of Mendel's laws of heredity might have gone unrecognized. Vital stains, dyes that can be used on living material, were first used in 1886 and have been greatly refined since then.

Improvement of the methodology of morphology has not been particularly rapid, even though satisfactory techniques for histology, anatomy, and cytology have been developed. The embedding of material in paraffin wax, the development of the rotary microtome for slicing very thin sections of tissue for microscope viewing, and the development of stain techniques are refinements of previously known methods. The invention of the phase microscope made possible the study of unfixed and unstained living material—hopefully nearer its natural state. The development of the electron microscope, however, has provided the plant morphologist with a new dimension of magnification of the structure of plant cells and tissues. The fine structure of the cell and of its components, such as mitochondria and the Golgi apparatus, have come under intensive study. Knowledge of the fine structure of plant cells has enabled investigators to determine the sites of important biochemical activities, especially those involved in the transfer of energy during photosynthesis and respiration. The scanning electron microscope, a relatively recent development, provides a three-dimensional image of surface structures at very great magnifications.

For experimental research on the morphogenesis of plants, isolated organs in their embryonic stage, clumps of cells, or even individual cells are grown. One of the most interesting techniques developed thus far permits the growing of plant tissue of higher plants as single cells; aeration and continuous agitation keep the cells suspended in the liquid culture medium.

Physiological Aspects

Plant physiology and plant biochemistry are the most technical areas of botany; most major advances in physiology also reflect the development of either a new technique or the dramatic refinement of an earlier one to give a new degree of precision. Fortunately, the methodology of measurement has been vastly improved in recent decades, largely through the development of various electronic devices. The phytotron at the California Institute of Technology represents the first serious attempt to control the environment of living plants on a relatively large scale; much important information has been gained concerning the effects on plants of day length and night length and the effects on growth, flowering, and fruiting of varying night temperatures. Critical measurements of other plant functions have also been obtained.

Certain complex biochemical processes, such as photosynthesis and respiration, have been studied stepwise by immobilizing the process through the use of extreme cold or biochemical inhibitors and by analyzing the enzymatic activity of specific cell contents after spinning cells at very high speeds in a centrifuge. The pathways of energy transfer from molecule to molecule during photosynthesis and respiration have been determined by biophysical methods, especially those utilizing radioactive isotopes.

An investigation of the natural metabolic products of plants requires, in general, certain standard biochemical techniques—e.g., gas and paper chromatography, electrophoresis, and various kinds of spectroscopy, including infrared, ultraviolet, and nuclear magnetic resonance. Useful information on the structure of the extremely large cellulose molecule has been provided by X-ray crystallography.

Ecological Aspects

When plant ecology first emerged as a subscience of botany, it was largely descriptive. Today, however, it has become a common meeting ground for all the plant sciences, as well as for other sciences. In addition, it has become much more quantitative. As a result, the tools and methods of plant ecologists are those available for measuring the intensity of the environmental factors that impinge on the plant and the reaction of the plant to these factors. The extent of the variability of many physical factors must be measured. The integration and reporting of such measurements, which cannot be regarded as constant, may therefore conceal some of the most dynamic and significant aspects of the environment and the responses of the plant to them. Because the physical environment is a complex of biological and physical components, it is measured by biophysical tools. The development of electronic measuring and recording devices has been crucial for a better understanding of the dynamics of the environment. Such devices, however, produce so much information that computer techniques must be used to reduce the data to meaningful results.

The ecologist might be concerned primarily with measuring the effect of the external environment on a plant and could adapt the methodology of the plant physiologist to field conditions.

The plant community ecologist is concerned with both the relation of different kinds of plants to each other and the nature and constitution of their association in natural communities. One widely used technique in this respect is to count the various kinds of plants within a standard area in order to determine such factors as the percentage of ground cover, dominance of species, aggressiveness, and other characteristics of the community. In general, the community ecologist has relatively few quantitative factors to measure, which nevertheless gives extremely useful results and some degree of predictability.

Some ecologists are most concerned with the inner environment of the plant and the way in which it reacts to the external environment. This approach, which is essentially physiological and bio-chemical, is useful for determining energy flow in ecosystems. The physiological ecologist is also concerned with evaluating the adaptations that certain plants have made toward survival in a hostile environment.

In summary, the techniques and methodology of plant ecology are as diverse and as varied as the large number of sciences that are drawn upon by ecologists. Completely new techniques, although few, are important; among them are techniques for measuring the amount of radioactive carbon-14 in plant deposits up to 50,000 years old. The most important new method in plant ecology is the rapidly growing use of computer techniques for handling vast amounts of data. Furthermore, modern digital computers can be used to simulate simple ecosystems and to analyze real ones.

Taxonomic Aspects

Experimental research under controlled conditions, made possible by botanical gardens and their ranges of greenhouses and controlled environmental chambers, has become an integral part of the methodology of modern plant taxonomy.

A second major tool of the taxonomist is the herbarium, a reference collection consisting of carefully selected and dried plants attached to paper sheets of a standard size and filed in a systematic way so that they may be easily retrieved for examination. Each specimen is a reference point representing the features of one plant of a certain species; it lasts indefinitely if properly cared for, and, if the species becomes extinct in nature—as thousands have—it remains the only record of the plant's former existence. The library is also an essential reference resource for descriptions and illustrations of plants that may not be represented in a particular herbarium.

Herbarium sheet: Herbarium sheet of a white tufted evening
primrose (Oenothera caespitosa variety marginata).

One of the earliest methods of the taxonomist, the study of living plants in the field, has benefited greatly by fast and easy methods of transportation. Botanists may carry on fieldwork in any part of the world and make detailed studies of the exact environmental conditions under which each species grows.

Many new approaches have been applied to the elucidation of problems in systematic botany. The transmission electron microscope and the scanning electron microscope have added to the knowledge of plant morphology, upon which classical taxonomy so much depends.

Allergen: Scanning electron microscopic
image of pollen from various common plants.

Refined methods for cytological and genetical studies of plants have given the taxonomist new insights into the origin of the great diversity among plants, especially the mechanisms by which new species arise and by which they then maintain their individuality in nature. From such studies have arisen further methods and also the subdisciplines of cytotaxonomy, cytogenetics, and population genetics.

Phytochemistry, or the chemistry of plants, one of the early subdivisions of organic chemistry, has been of great importance in the identification of plant substances of medicinal importance. With the development of new phytochemical methods, new information has become available for use in conjunction with plant taxonomy. Thus has arisen the modern field of chemotaxonomy, or biochemical systematics. Each species tends to differ to some degree from every other species, even in the same genus, in the biochemistry of its natural metabolic products. Sometimes the difference is subtle and difficult to determine; sometimes it is obvious and easily perceptible. With new analytical techniques, a large number of individual compounds from one plant can be identified quickly and with certainty. Such information is extremely useful in adding confirmatory or supplemental evidence of an objective and quantitative nature. An interesting by-product of chemical plant taxonomy has resulted in understanding better the restriction of certain insects to specific plants.

Computer techniques have been applied to plant taxonomy to develop a new field, numerical taxonomy, or taximetrics, by which relationships between plant species or those within groups of species are determined quantitatively and depicted graphically. Another method measures the degree of molecular similarity of deoxyribonucleic acid (DNA) molecules in different plants. By this procedure it should be possible to determine the natural taxonomic relationships (phylogeny) among different plants and plant groups by determining the extent of the relationship of their DNA: closely related plants will have more similarities in their DNA than will unrelated ones.

Animal Biology

Zoology is the branch of biology that studies the members of the animal kingdom and animal life in general. It includes both the inquiry into individual animals and their constituent parts, even to the molecular level, and the inquiry into animal populations, entire faunas, and the relationships of animals to each other, to plants, and to the nonliving environment. Though this wide range of studies results in some isolation of specialties within zoology, the conceptual integration in the contemporary study of living things that has occurred in recent years emphasizes the structural and functional unity of life rather than its diversity.

Prehistoric man's survival as a hunter defined his relation to other animals, which were a source of food and danger. As man's cultural heritage developed, animals were variously incorporated into man's folklore and philosophical awareness as fellow living creatures. Domestication of animals forced man to take a systematic and measured view of animal life, especially after urbanization necessitated a constant and large supply of animal products.

Study of animal life by the ancient Greeks became more rational, if not yet scientific, in the modern sense, after the cause of disease—until then thought to be demons—was postulated by Hippocrates to result from a lack of harmonious functioning of body parts. The systematic study of animals was encouraged by Aristotle's extensive descriptions of living things, his work reflecting the Greek concept of order in nature and attributing to nature an idealized rigidity.

In Roman times Pliny brought together in 37 volumes a treatise, Historia naturalis, that was an encyclopaedic compilation of both myth and fact regarding celestial bodies, geography, animals and plants, metals, and stone. Volumes VII to XI concern zoology; volume VIII, which deals with the land animals, begins with the largest one, the elephant. Although Pliny's approach was naïve, his scholarly effort had a profound and lasting influence as an authoritative work.

Zoology continued in the Aristotelian tradition for many centuries in the Mediterranean region and by the Middle Ages, in Europe, it had accumulated considerable folklore, superstition, and moral symbolisms, which were added to otherwise objective information about animals. Gradually, much of this misinformation was sifted out: naturalists became more critical as they compared directly observed animal life in Europe with that described in ancient texts. The use of the printing press in the 15th century made possible an accurate transmission of information. Moreover, mechanistic views of life processes (i.e., that physical processes depending on cause and effect can apply to animate forms) provided a hopeful method for analyzing animal functions; for example, the mechanics of hydraulic systems were part of William Harvey's argument for the circulation of the blood—although Harvey remained thoroughly Aristotelian in outlook. In the 18th century, zoology passed through reforms provided by both the system of nomenclature of Carolus Linnaeus and the comprehensive works on natural history by Georges-Louis Leclerc de Buffon; to these were added the contributions to comparative anatomy by Georges Cuvier in the early 19th century.

Physiological functions, such as digestion, excretion, and respiration, were easily observed in many animals, though they were not as critically analyzed as was blood circulation.

Following the introduction of the word cell in the 17th century and microscopic observation of these structures throughout the 18th century, the cell was incisively defined as the common structural unit of living things in 1839 by two Germans: Matthias Schleiden and Theodor Schwann. In the meanwhile, as the science of chemistry developed, it was inevitably extended to an analysis of animate systems. In the middle of the 18th century the French physicist René Antoine Ferchault de Réaumer demonstrated that the fermenting action of stomach juices is a chemical process. And in the mid-19th century the French physician and physiologist Claude Bernard drew upon both the cell theory and knowledge of chemistry to develop the concept of the stability of the internal bodily environment, now called homeostasis.

The cell concept influenced many biological disciplines, including that of embryology, in which cells are important in determining the way in which a fertilized egg develops into a new organism. The unfolding of these events—called epigenesis by Harvey—was described by various workers, notably the German-trained comparative embryologist Karl von Baer, who was the first to observe a mammalian egg within an ovary. Another German-trained embryologist, Christian Heinrich Pander, introduced in 1817 the concept of germ, or primordial, tissue layers into embryology.

In the latter part of the 19th century, improved microscopy and better staining techniques using aniline dyes, such as hematoxylin, provided further impetus to the study of internal cellular structure.

By this time Darwin had made necessary a complete revision of man's view of nature with his theory that biological changes in species occur through the process of natural selection. The theory of evolution—that organisms are continuously evolving into highly adapted forms—required the rejection of the static view that all species are especially created and upset the Linnaean concept of species types. Darwin recognized that the principles of heredity must be known to understand how evolution works; but, even though the concept of hereditary factors had by then been formulated by Mendel, Darwin never heard of his work, which was essentially lost until its rediscovery in 1900.

Genetics has developed in the 20th century and now is essential to many diverse biological disciplines. The discovery of the gene as a controlling hereditary factor for all forms of life has been a major accomplishment of modern biology. There has also emerged clearer understanding of the interaction of organisms with their environment. Such ecological studies help not only to show the interdependence of the three great groups of organisms—plants, as producers; animals, as consumers; and fungi and many bacteria, as decomposers—but they also provide information essential to man's control of the environment and, ultimately, to his survival on Earth. Closely related to this study of ecology are inquiries into animal behavior, or ethology. Such studies are often cross disciplinary in that ecology, physiology, genetics, development, and evolution are combined as man attempts to understand why an organism behaves as it does. This approach now receives substantial attention because it seems to provide useful insight into man's biological heritage—that is, the historical origin of man from nonhuman forms.

The emergence of animal biology has had two particular effects on classical zoology. First, and somewhat paradoxically, there has been a reduced emphasis on zoology as a distinct subject of scientific study; for example, workers think of themselves as geneticists, ecologists, or physiologists who study animal rather than plant material. They often choose a problem congenial to their intellectual tastes, regarding the organism used as important only to the extent that it provides favourable experimental material. Current emphasis is, therefore, slanted toward the solution of general biological problems; contemporary zoology thus is to a great extent the sum total of that work done by biologists pursuing research on animal material.

Second, there is an increasing emphasis on a conceptual approach to the life sciences. This has resulted from the concepts that emerged in the late 19th and early 20th centuries: the cell theory; natural selection and evolution; the constancy of the internal environment; the basic similarity of genetic material in all living organisms; and the flow of matter and energy through ecosystems. The lives of microbes, plants, and animals now are approached using theoretical models as guides rather than by following the often restricted empiricism of earlier times. This is particularly true in molecular studies, in which the integration of biology with chemistry allows the techniques and quantitative emphases of the physical sciences to be used effectively to analyze living systems.

Areas of Study

Although it is still useful to recognize many disciplines in animal biology—e.g., anatomy or morphology; biochemistry and molecular biology; cell biology; developmental studies (embryology); ecology; ethology; evolution; genetics; physiology; and systematics—the research frontiers occur as often at the interfaces of two or more of these areas as within any given one.

Anatomy or Morphology

Descriptions of external form and internal organization are among the earliest records available regarding the systematic study of animals. Aristotle was an indefatigable collector and dissector of animals. He found differing degrees of structural complexity, which he described with regard to ways of living, habits, and body parts. Although Aristotle had no formal system of classification, it is apparent that he viewed animals as arranged from the simplest to the most complex in an ascending series. Since man was even more complex than animals and, moreover, possessed a rational faculty, he therefore occupied the highest position and a special category. This hierarchical perception of the animate world proved to be useful in every century to the present, except that in the modern view there is no such "scale of nature," and there is change in time by evolution from the simple to the complex.

After the time of Aristotle, Mediterranean science was centred at Alexandria, where the study of anatomy, particularly the central nervous system, flourished and, in fact, first became recognized as a discipline. Galen studied anatomy at Alexandria in the 2nd century and later dissected many animals. Much later, the contributions of the Renaissance anatomist Andreas Vesalius, though made in the context of medicine, as were those of Galen, stimulated to a great extent the rise of comparative anatomy. During the latter part of the 15th century and throughout the 16th century, there was a strong tradition in anatomy; important similarities were observed in the anatomy of different animals, and many illustrated books were published to record these observations.

But anatomy remained a purely descriptive science until the advent of functional considerations in which the correlation between structure and function was consciously investigated; as by French biologists Buffon and Cuvier. Cuvier cogently argued that a trained naturalist could deduce from one suitably chosen part of an animal's body the complete set of adaptations that characterized the organism. Because it was obvious that organisms with similar parts pursue similar habits, they were placed together in a system of classification. Cuvier pursued this viewpoint, which he called the theory of correlations, in a somewhat dogmatic manner and placed himself in opposition to the romantic natural philosophers, such as the German intellectual Johann Wolfgang von Goethe, who saw a tendency to ideal types in animal form. The tension between these schools of thought—adaptation as the consequence of necessary bodily functions and adaptation as an expression of a perfecting principle in nature—runs as a leitmotiv through much of biology, with overtones extending into the early 20th century.

The twin concepts of homology (similarity of origin) and analogy (similarity of appearance), in relation to structure, are the creation of the 19th-century British anatomist Richard Owen. Although they antedate the Darwinian view of evolution, the anatomical data on which they were based became, largely as a result of the work of the German comparative anatomist Carl Gegenbaur, important evidence in favour of evolutionary change, despite Owen's steady unwillingness to accept the view of diversification of life from a common origin.

In summary, anatomy moved from a purely descriptive phase as an adjunct to classificatory studies, into a partnership with studies of function and became, in the 19th century, a major contributor to the concept of evolution.

Taxonomy or Systematics

Not until the work of Carolus Linnaeus did the variety of life receive a widely accepted systematic treatment. Linnaeus strove for a "natural method of arrangement," one that is now recognizable as an intuitive grasp of homologous relationships, reflecting evolutionary descent from a common ancestor; however, the natural method of arrangement sought by Linnaeus was more akin to the tenets of idealized morphology because he wanted to define a "type" form as epitomizing a species.

Linnaeus, Carolus: Systema Naturae.

It was in the nomenclatorial aspect of classification that Linnaeus created a revolutionary advance with the introduction of a Latin binomial system: each species received a Latin name, which was not influenced by local names and which invoked the authority of Latin as a language common to the learned people of that day. The Latin name has two parts. The first word in the Latin name for the common chimpanzee, Pan troglodytes, for example, indicates the larger category, or genus, to which chimpanzees belong; the second word is the name of the species within the genus. In addition to species and genera, Linnaeus also recognized other classificatory groups, or taxa (singular taxon), which are still used; namely, order, class, and kingdom, to which have been added family (between genus and order) and phylum (between class and kingdom). Each of these can be divided further by the appropriate prefix of sub- or super-, as in subfamily or superclass. Linnaeus' great work, the Systema naturae, went through 12 editions during his lifetime; the 13th, and final, edition appeared posthumously. Although his treatment of the diversity of living things has been expanded in detail, revised in terms of taxonomic categories, and corrected in the light of continuing work—for example, Linnaeus treated whales as fish—it still sets the style and method, even to the use of Latin names, for contemporary nomenclatorial work.

Linnaeus sought a natural method of arrangement, but he actually defined types of species on the basis of idealized morphology. The greatest change from Linnaeus' outlook is reflected in the phrase "the new systematics," which was introduced in the 20th century and through which an explicit effort is made to have taxonomic schemes reflect evolutionary history. The basic unit of classification, the species, is also the basic unit of evolution—i.e., a population of actually or potentially interbreeding individuals. Such a population shares, through interbreeding, its genetic resources. In so doing, it creates the gene pool—its total genetic material—that determines the biological resources of the species and on which natural selection continuously acts. This approach has guided work on classifying animals away from somewhat arbitrary categorization of new species to that of recreating evolutionary history (phylogeny) and incorporating it in the system of classification. Modern taxonomists or systematists, therefore, are among the foremost students of evolution.

Physiology

The practical consequences of physiology have always been an unavoidable human concern, in both medicine and animal husbandry. Inevitably, from Hippocrates to the present, practical knowledge of human bodily function has accumulated along with that of domestic animals and plants. This knowledge has been expanded, especially since the early 1800s, by experimental work on animals in general, a study known as comparative physiology. The experimental dimension had wide applications following Harvey's demonstration of the circulation of blood. From then on, medical physiology developed rapidly; notable texts appeared, such as Albrecht von Haller's eight-volume work Elementa Physiologiae Corporis Humani (Elements of Human Physiology), which had a medical emphasis. Toward the end of the 18th century the influence of chemistry on physiology became pronounced through Antoine Lavoisier's brilliant analysis of respiration as a form of combustion. This French chemist not only determined that oxygen was consumed by living systems but also opened the way to further inquiry into the energetics of living systems. His studies further strengthened the mechanistic view, which holds that the same natural laws govern both the inanimate and the animate realms.

Physiological principles achieved new levels of sophistication and comprehensiveness with Bernard's concept of constancy of the internal environment, the point being that only under certain constantly maintained conditions is there optimal bodily function. His rational and incisive insights were augmented by concurrent developments in Germany, where Johannes Müller explored the comparative aspects of animal function and anatomy, and Justus von Liebig and Carl Ludwig applied chemical and physical methods, respectively, to the solution of physiological problems. As a result, many useful techniques were advanced—e.g., means for precise measurement of muscular action and changes in blood pressure and means for defining the nature of body fluids.

By this time the organ systems—circulatory, digestive, endocrine, excretory, integumentary, muscular, nervous, reproductive, respiratory, and skeletal—had been defined, both anatomically and functionally, and research efforts were focussed on understanding these systems in cellular and chemical terms, an emphasis that continues to the present and has resulted in specialties in cell physiology and physiological chemistry. General categories of research now deal with the transportation of materials across membranes; the metabolism of cells, including synthesis and breakdown of molecules; and the regulation of these processes.

Interest has also increased in the most complex of physiological systems, the nervous system. Much comparative work has been done by utilizing animals with structures especially amenable to various experimental techniques; for example, the large nerves in squids have been extensively studied in terms of the transmission of nerve impulses, and insect and crustacean eyes have yielded significant information on patterns of sensory inputs. Most of this work is closely associated with studies on animal orientation and behavior. Although the contemporary physiologist often studies functional problems at the molecular and cellular levels, he is also aware of the need to integrate cellular studies into the many-faceted functions of the total organism.

Embryology or Developmental Studies

Embryonic growth and differentiation of parts have been major biological problems since ancient times. A 17th-century explanation of development assumed that the adult existed as a miniature—a homunculus—in the microscopic material that initiates the embryo. But in 1759 the German physician Caspar Friedrick Wolff firmly introduced into biology the interpretation that undifferentiated materials gradually become specialized, in an orderly way, into adult structures. Although this epigenetic process is now accepted as characterizing the general nature of development in both plants and animals, many questions remain to be solved. The French physician Marie François Xavier Bichat declared in 1801 that differentiating parts consist of various components called tissues; with the subsequent statement of the cell theory, tissues were resolved into their cellular constituents. The idea of epigenetic change and the identification of structural components made possible a new interpretation of differentiation. It was demonstrated that the egg gives rise to three essential germ layers out of which specialized organs, with their tissues, subsequently emerge. Then, following his own discovery of the mammalian ovum, von Baer in 1828 usefully applied this information when he surveyed the development of various members of the vertebrate groups. At this point, embryology, as it is now recognized, emerged as a distinct subject.

The concept of cellular organization had an effect on embryology that continues to the present day. In the 19th century, cellular mechanisms were considered essentially to be the basis for growth, differentiation, and morphogenesis, or molding of parts. The distribution of the newly formed cells of the rapidly dividing zygote (fertilized egg) was precisely followed to provide detailed accounts not only of the time and mode of germ layer formation but also of the contribution of these layers to the differentiation of tissues and organs. Such descriptive information provided the background for experimental work aimed at elucidating the role of chromosomes and other cellular constituents in differentiation. About 1895, before the formulation of the chromosomal theory of heredity, Theodor Boveri demonstrated that chromosomes show continuity from one cell generation to the next. In fact, biologists soon concluded that in all cells arising from a fertilized egg, half the chromosomes are of maternal and half of paternal origin. The discovery of the constant transmission of the original chromosomal endowment to all cells of the body served to deepen the mystery surrounding the factors that determine cellular differentiation.

The present view is that differential activity of genes is the basis for cellular and tissue differentiation; that is, although the cells of a multicellular body contain the same genetic information, different genes are active in different cells. The result is the formation of various gene products, which regulate the functional and structural differentiation of cells. The actual mechanism involved in the inactivation of certain genes and the activation of others, however, has not yet been established. That cells can move extensively throughout the embryo and selectively adhere to other cells, thus starting tissue aggregations, also contributes to development as does the fate of cells—i.e., certain ones continue to multiply, others stop, and some die.

Research methods in embryology now exploit many experimental situations: both unicellular and multicellular forms; regeneration (replacement of lost parts) and normal development; and growth of tissues outside and inside the host. Hence, the processes of development can be studied with material other than embryos; and the study of embryology has become incorporated into the more inclusive subdiscipline of developmental biology.

Evolutionism

Darwin was not the first to speculate that organisms can change from generation to generation and so evolve, but he was the first to propose a mechanism by which the changes are accumulated. He proposed that heritable variations occur in conjunction with a never-ending competition for survival and that the variations favouring survival are automatically preserved. In time, therefore, the continued accumulation of variations results in the emergence of new forms. Because the variations that are preserved relate to survival, the survivors are highly adapted to their environment. To this process Darwin gave the apt name natural selection.

Many of Darwin's predecessors, notably Jean-Baptiste Lamarck, were willing to accept the idea of species variation, even though to do so meant denying the doctrine of special creation and the static-type species of Linnaeus. But they argued that some idealized perfecting principle, expressed through the habits of an organism, was the basis of variation. The contrast between the romanticism of Lamarck and the objective analysis of Darwin clearly reveals the type of revolution provoked by the concept of natural selection. Although mechanistic explanations had long been available to biologists—forming, for example, part of Harvey's explanation of blood circulation—they did not pervade the total structure of biological thinking until the advent of Darwinism.

There were two immediate consequences of Darwin's viewpoints. One has involved a reappraisal of all subject areas of biology; reinterpretations of morphology and embryology are good examples. The comparative anatomy of the British anatomist Owen became a cornerstone of the evidence for evolution, and German anatomists provided the basis for the comment that evolutionary thinking was born in England but gained its home in Germany. The reinterpretation of morphology carried over into the study of fossil forms, as paleontologists sought and found evidence of gradual change in their study of fossils. But some workers, although accepting evolution in principle, could not easily interpret the changes in terms of natural selection. The German paleontologist Otto Schindewolf, for example, found in shelled mollusks called ammonites evidence of progressive complexity and subsequent simplification of forms. The American paleontologist George Gaylord Simpson, however, has been a consistent interpreter of vertebrate fossils by Darwinian selection. Embryology was seen in an evolutionary light when the German zoologist Ernst Haeckel proposed that the epigenetic sequence of embryonic development (ontogeny) repeated its evolutionary history (phylogeny). Thus, the presence of gill clefts in the mammalian embryo and also in less highly evolved vertebrates can be understood as a remnant of a common ancestor.

The other consequence of Darwinism—to make more explicit the origin and nature of heritable variations and the action of natural selection on them—depended on the emergence of the following: genetics and the elucidation of the rules of Mendelian inheritance; the concept of the gene as the unit of inheritance; and the nature of gene mutation. The development of these ideas provided the basis for the genetics of natural populations.

The subject of population genetics began with the Mendelian laws of inheritance and now takes into account selection, mutation, migration (movement into and out of a given population), breeding patterns, and population size. These factors affect the genetic makeup of a group of organisms that either interbreed or have the potential to do so; i.e., a species. Accurate appraisal of these factors allows precise predictions regarding the content of a given gene pool over significant periods of evolutionary time. From work involving population genetics has come the realization, eloquently documented by two contemporary American evolutionists, Theodosius Dobzhansky and Ernst Mayer, that the species is the basic unit of evolution. The process of speciation occurs as a gene pool breaks up to form isolated gene pools. When selection pressures similar to those of the original gene pool persist in the new gene pools, similar functions and the similar structures on which they depend also persist. When selection pressures differ, however, differences arise. Thus, the process of speciation through natural selection preserves the evolutionary history of a species. The record may be discerned not only in the gross, or macroscopic, anatomy of organisms but also in their cellular structure and molecular organization. Significant work now is carried out, for example, on the homologies of the nucleic acids and proteins of different species.

Genetics

The problem of heredity had been the subject of careful study before its definitive analysis by Mendel. As with Darwin's predecessors, those of Mendel tended to idealize and interpret all inherited traits as being transmitted through the blood or as determined by various "humors" or other vague entities in animal organisms. When studying plants, Mendel was able to free himself of anthropomorphic and holistic explanations. By studying seven carefully defined pairs of characteristics—e.g., tall and short plants; red and white flowers, etc.—as they were transmitted through as many as three successive generations, he was able to establish patterns of inheritance that apply to all sexually reproducing forms. Darwin, who was searching for an explanation of inheritance, apparently never saw Mendel's work, which was published in 1866 in the obscure journal of his local natural history society; it was simultaneously rediscovered in 1900 by three different European geneticists.

Further progress in genetics was made early in the 20th century, when it was realized that heredity factors are found on chromosomes. The term gene was coined for these factors. Studies by the American geneticist Thomas Hunt Morgan on the fruit fly (Drosophila), moved animal genetics to the forefront of genetic research. The work of Morgan and his students established such major concepts as the linear array of genes on chromosomes; the exchange of parts between chromosomes; and the interaction of genes in determining traits, including sexual differences. In 1927 one of Morgan's former students, Hermann Muller, used X rays to induce the mutations (changes in genes) in the fruit fly, thereby opening the door to major studies on the nature of variation.

Meanwhile, other organisms were being used for genetic studies, most notably fungi and bacteria. The results of this work provided insights into animal genetics just as principles initially obtained from animal genetics provided insight into botanical and microbial forms. Work continues not only on the genetics of humans, domestic animals, and plants but also on the control of development through the orderly regulation of gene action in different cells and tissues.

Cellular and Molecular Biology

Although the cell was recognized as the basic unit of life early in the 19th century, its most exciting period of inquiry has probably occurred since the 1940s. The new techniques developed since that time, notably the perfection of the electron microscope and the tools of biochemistry, have changed the cytological studies of the 19th and early 20th centuries from a largely descriptive inquiry, dependent on the light microscope, into a dynamic, molecularly oriented inquiry into fundamental life processes.

The so-called cell theory, which was enunciated about 1838, was never actually a theory. As Edmund Beecher Wilson, the noted American cytologist, stated in his great work, The Cell,

> By force of habit we still continue to speak of the cell 'theory' but it is a theory only in name. In substance it is a comprehensive general statement of fact and as such stands today beside the evolution theory among the foundationstones of modern biology.

More precisely, the cell doctrine was an inductive generalization based on the microscopial examination of certain plant and animal species.

Rudolf Virchow, a German medical officer specializing in cellular pathology, first expressed the fundamental dictum regarding cells in his phrase omnis cellula e cellula (all cells from cells). For cellular reproduction is the ultimate basis of the continuity of life; the cell is not only the basic structural unit of life but also the basic physiological and reproductive unit. All areas of biology were affected by the new perspective afforded by the principle of cellular organization. Especially in conjunction with embryology was the study of the cell most prominent in animal biology. The continuity of cellular generations by reproduction also had implications for genetics. It is little wonder, then, that the full title of Wilson's survey of cytology at the turn of the century was The Cell: Its Role in Development and Heredity.

The study of the cell nucleus, its chromosomes, and their behavior served as the basis for understanding the regular distribution of genetic material during both sexual and asexual reproduction. This orderly behavior of the nucleus made it appear to dominate the life of the cell, for by contrast the components of the rest of the cell appeared to be randomly distributed.

The biochemical study of life had helped in the characterization of the major molecules of living systems—proteins, nucleic acids, fats, and carbohydrates—and in the understanding of metabolic processes. That nucleic acids are a distinctive feature of the nucleus was recognized after their discovery by the Swiss biochemist Johann Friedrich Miescher in 1869. In 1944 a group of American bacteriologists, led by Oswald T. Avery, published work on the causative agent of pneumonia in mice (a bacterium) that culminated in the demonstration that deoxyribonucleic acid (DNA) is the chemical basis of heredity. Discrete segments of DNA correspond to genes, or Mendel's hereditary factors. Proteins were discovered to be especially important for their role in determining cell structure and in controlling chemical reactions.

The advent of techniques for isolating and characterizing proteins and nucleic acids now allows a molecular approach to essentially all biological problems—from the appearance of new gene products in normal development or under pathological conditions to a monitoring of changes in and between nerve cells during the transmission of nerve impulses.

Ecology

The harmony that Linnaeus found in nature, which redounded to the glory and wisdom of a Judaeo-Christian god, was the 18th-century counterpart of the balanced interaction now studied by ecologists. Linnaeus recognized that plants are adapted to the regions in which they grow, that insects play a role in flower pollination, and that certain birds prey on insects and are in turn eaten by other birds. This realization implies, in contemporary terms, the flow of matter and energy in a definable direction through any natural assemblage of plants, animals, and microorganisms. Such an assemblage, termed an ecosystem, starts with the plants, which are designated as producers because they maintain and reproduce themselves at the expense of energy from sunlight and inorganic materials taken from the nonliving environment around them (earth, air, and water). Animals are called consumers because they ingest plant material or other animals that feed on plants, using the energy stored in this food to sustain themselves. Lastly, the organisms known as decomposers, mostly fungi and bacteria, break down plant and animal material and return it to the environment in a form that can be used again by plants in a constantly renewed cycle.

The term ecology, first formulated by Haeckel in the latter part of the 19th century as "oecology", referred to the dwelling place of organisms in nature. In the 1890s various European and U.S. scientists laid the foundations for modern work through studies of natural ecosystems and the populations of organisms contained within them.

Animal ecology, the study of consumers and their interactions with the environment, is very complex; attempts to study it usually focus on one particular aspect. Some studies, for example, involve the challenge of the environment to individuals with special adaptations (e.g., water conservation in desert animals); others may involve the role of one species in its ecosystem or the ecosystem itself. Food-chain sequences have been determined for various ecosystems, and the efficiency of the transfer of energy and matter within them has been calculated so that their capacity is known; that is, productivity in terms of numbers of organisms or weight of living matter at a specific level in the food chain can be accurately determined.

In spite of advances in understanding animal ecology, this subject area of zoology does not yet have the major unifying theoretical principles found in genetics (gene theory) or evolution (natural selection).

Ethology

The study of animal behavior (ethology) is largely a 20th-century phenomenon and is exclusively a zoological discipline. Only animals have nervous systems, with their implications for perception, coordination, orientation, learning, and memory. Not until the end of the 19th century did animal behavior become free from anthropocentric interests and assume an importance in its own right. The British behaviorist C. Lloyd Morgan was probably most influential with his emphasis on parsimonious explanations—i.e., that the explanation "which stands lower in the psychological scale" must be invoked first. This principle is exemplified in the American Herbert Spencer Jennings' pioneering work in 1906 on The Behavior of Lower Organisms.

The study of animal behavior now includes many diverse topics, ranging from swimming patterns of protozoans to socialization and communication among the great apes. Many disparate hypotheses have been proposed in an attempt to explain the variety of behavioral patterns found in animals. They focus on the mechanisms that stimulate courtship in reproductive behavior of such diverse groups as spiders, crabs, and domestic fowl; and on whole life histories, starting from the special attachment of newly born ducks and goats to their actual mothers or to surrogate (substitute) mothers. The latter phenomenon, called imprinting, has been intensively studied by the Austrian ethologist Konrad Lorenz. Physiologically oriented behavior now receives much attention; studies range from work on conditioned reflexes to the orientation of crustaceans and the location and communication of food among bees; such diversity of material is one measure of the somewhat diffuse but exciting current state of these studies.

General Trends

Zoology has become animal biology—that is, the life sciences display a new unity, one that is founded on the common basis of all life, on the gene pool–species organization of organisms, and on the obligatory interacting of the components of ecosystems. Even as regards the specialized features of animals—involving physiology, development, or behavior—the current emphasis is on elucidating the broad biological principles that identify animals as one aspect of nature. Zoology has thus given up its exclusive emphasis on animals—an emphasis maintained from Aristotle's time well into the 19th century—in favour of a broader view of life. The successes in applying physical and chemical ideas and techniques to life processes have not only unified the life sciences but have also created bridges to other sciences in a way only dimly foreseen by earlier workers. The practical and theoretical consequences of this trend have just begun to be realized.

Methods in Zoology

Because the study of animals may be concentrated on widely different topics, such as ecosystems and their constituent populations, organisms, cells, and chemical reactions, specific techniques are needed for each kind of investigation. The emphasis on the molecular basis of genetics, development, physiology, behavior, and ecology has placed increasing importance on those techniques involving cells and their many components. Microscopy, therefore, is a necessary technique in zoology, as are certain physicochemical methods for isolating and characterizing molecules. Computer technology also has a special role in the analysis of animal life. These newer techniques are used in addition to the many classical ones—measurement and experimentation at the tissue, organ, organ system, and organismic levels.

Microscopy

In addition to continuous improvements in the techniques of staining cells, so that their components can be seen clearly, the light used in microscopy can now be manipulated to make visible certain structures in living cells that are otherwise undetectable. The ability to observe living cells is an advantage of light microscopes over electron microscopes; the latter require the cells to be in an environment that kills them. The particular advantage of the electron microscope, however, is its great powers of magnification. Theoretically, it can resolve single atoms; in biology, however, magnifications of lesser magnitude are most useful in determining the nature of structures lying between whole cells and their constituent molecules.

Separation and Purification Techniques

The characterization of components of cellular systems is necessary for biochemical studies. The specific molecular composition of cellular organelles, for example, affects their shape and density (mass per unit volume); as a result, cellular components settle at different rates (and thus can be separated) when they are spun in a centrifuge.

Other methods of purification rely on other physical properties. Molecules vary in their affinity for the positive or negative pole of an electrical field. Migration to or away from these poles, therefore, occurs at different rates for different molecules and allows their separation; the process is called electrophoresis. The separation of molecules by liquid solvents exploits the fact that the molecules differ in their solubility, and hence they migrate to various degrees as a solvent flows past them. This process, known as chromatography because of the colour used to identify the position of the migrating materials, yields samples of extraordinarily high purity.

Radioactive Tracers

Radioactive compounds are especially useful in biochemical studies involving metabolic pathways of synthesis and degradation. Radioactive compounds are incorporated into cells in the same way as their nonradioactive counterparts. These compounds provide information on the sites of specific metabolic activities within cells and insights into the fates of these compounds in both organisms and the ecosystem.

Computers

Computers process information using their own general language, which is able to complete calculations as complex and diverse as statistical analyses and determinations of enzymatically controlled reaction rates. Computers with access to extensive data files can select information associated with a specific problem and display it to aid the researcher in formulating possible solutions. They help perform routine examinations such as scanning chromosome preparations in order to identify abnormalities in number or shape. Test organisms can be electronically monitored with computers, so that adjustments can be made during experiments; this procedure improves the quality of the data and allows experimental situations to be fully exploited. Computer simulation is important in analyzing complex problems; as many as 100 variables, for example, are involved in the management of salmon fisheries. Simulation makes possible the development of models that approach the complexities of conditions in nature, a procedure of great value in studying wildlife management and related ecological problems.

Applied Zoology

Animal-related industries produce food (meats and dairy products), hides, furs, wool, organic fertilizers, and miscellaneous chemical byproducts. There has been a dramatic increase in the productivity of animal husbandry since the 1870s, largely as a consequence of selective breeding and improved animal nutrition. The purpose of selective breeding is to develop livestock whose desirable traits have strong heritable components and can therefore be propagated. Heritable components are distinguished from environmental factors by determining the coefficient of heritability, which is defined as the ratio of variance in a gene-controlled character to total variance.

Another aspect of food production is the control of pests. The serious side effects of some chemical pesticides make extremely important the development of effective and safe control mechanisms. Animal food resources include commercial fishing. The development of shellfish resources and fisheries management (e.g., growth of fish in rice paddies in Asia) are important aspects of this industry.

References

- Caldwell R (2006). "Comparative Anatomy: Andreas Vesalius". University of California Museum of Paleontology. Archived from the original on 2010-11-23. Retrieved 2011-02-17

- Anatomy, science: britannica.com, Retrieved 31 March, 2019

- Gaucher EA, Kratzer JT, Randall RN (January 2010). "Deep phylogeny--how a tree can help characterize early life on Earth". Cold Spring Harbor Perspectives in Biology. 2 (1): a002238. Doi:10.1101/cshperspect.a002238. PMC 2827910. PMID 20182607

- Rosen J, Gothard LQ (2009). Encyclopedia of Physical Science. Infobase Publishing. P. 4 9. ISBN 978-0-8160-7011-4

- Sahai, Erik; Trepat, Xavier (July 2018). "Mesoscale physical principles of collective cell organization". Nature Physics. 14 (7): 671–682. Doi:10.1038/s41567-018-0194-9. ISSN 1745-2481

- Ettensohn C.A., Sweet H.C. (2000). Patterning the early sea urchin embryo. Curr. Top. Dev. Biol. Current Topics in Developmental Biology. 50. Pp. 1–44. Doi:10.1016/S0070-2153(00)50002-7. ISBN 9780121531508

- "Specialized literature". Islamic culture and medical arts. U.S. National Library of Medicine. Retrieved 24 September 2013

- Rowles, Graham D.; Bernard, Miriam (2013). Environmental Gerontology: Making Meaningful Places in Old Age. New York: Springer Publishing Company. P. 320. ISBN 978-0826108135

3
Genetics: The Study of Genes

The branch of biology which focuses on the study of genes, genetic variation and heredity in organisms is termed as genetics. Some of the major areas of study within this field are regulation of gene expression, extranuclear inheritance and genetic linkage. This chapter discusses in detail these key focus areas related to genetics.

Gene

Genes are the units of heredity in living organisms, are encoded in an organism's genetic material (DNA). They exert a central influence on the organism's physical aspects and are passed on to succeeding generations through the reproduction process. Genetic material can also be passed between unrelated individuals on viruses or through the process of transfection used in genetic engineering.

Common usage of the word "gene" reflects its meaning in molecular biology, namely the segments of DNA that cells transcribe into either RNA that is translated into proteins (DNA=>RNA=>protein) or RNA used for direct purposes (DNA=>RNA). The Sequence Ontology project, a consortium of several centers of genomic studies, defines a gene as: "A locatable region of genomic sequence, corresponding to a unit of inheritance, which is associated with regulatory regions, transcribed regions, and/or other functional sequence regions." The definition reflects the full complexity that has come to be associated with the term gene.

Genes encode the information necessary for constructing the multitude of proteins and RNA units needed to maintain an organism's existence, growth, action, and multiplication. Each gene that serves as the first step in protein formation is a region of DNA comprising a mixture of some sections (exons) that code for proteins, others (introns) that have no apparent function, and still others that define the beginning and end of the gene or the conditions in which the gene will be expressed or not expressed.

Although the human genome comprises roughly 25,000 genes carrying codes for proteins, each human cell has the potential of making about 100,000 different proteins. Further complexity lies in the additional 10,000 or so genes used for making RNA that directly serves such cellular functions as structure, catalysis, and regulation of gene expression. The proteins and RNAs all share in the tasks of maintaining the cell, one of which is the continual fine tuning of the exact selection of genes being expressed according to the cell's function and its continually changing environment.

The ongoing discovery of so much functional RNA in the cell, much of it related to the expression of genes, is taken be some as a sign that RNA may deserve a co-equal billing with DNA in terms of overall contribution to the cellular function.

Genes are of central importance to the physical aspect of a living organism: A person's eye color, the breed of a dog, the gender of a horse. Mouse DNA yields a mouse, not an elephant. However, the impact of genes is sometimes extrapolated to the view that genes control everything about human lives and destiny. This is the concept of genetic determinism whereby human behavior, intelligence, emotions and attitudes, and health are fixed by genetic makeup and thus unchangeable. Such a misconception has at times been used as a base for explaining away racial prejudices, addictions, and criminal behavior, and seeking solutions to social problems by turning to genetic engineering as the ultimate solution.

The more balanced and generally recognized view is that biological contributions to solving social problems must be sought through a biology that takes into account the influence of social and cultural factors in human physical development and behavior.

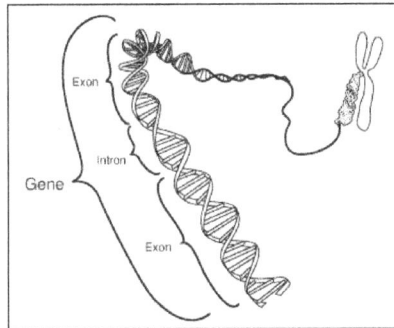

Diagram showing a greatly simplified gene in relation
to the double helix structure of DNA with intron
(non-coding) and exon (coding) portions labeled.

If the gene is active in a particular cell, it will be transcribed into RNA that serves either a direct function in the cell or as a template for making a protein. In eukaryotic organisms, making protein from RNA entails excising the intron regions and splicing together some of the exon regions to make the template for one or more different proteins depending on which of the RNA exon regions are spliced together.

Genetics

Genetics is the study of genes and inheritance in living organisms. This branch of science has a fascinating history, stretching from the 19th century when scientists began to study how organisms inherited traits from their parents, to the present day when we can read the "source code" of living things letter-by-letter.

Genetics started out with curiosity about why things are the way things are – why do children resemble one parent more than another? Why do some species resemble each other more closely than others?

It has evolved into an almost universal answer handbook for biology. By reading the "source code" or "blueprint" for an organism, scientists today are often able to pinpoint exactly where an organism came from, how it has changed over time, what diseases it might develop, and how its life processes are similar to or different from those of other organisms.

Impact of Genetics

The ability to read the "source code of life" has led to a revolution in the way we think about and classify organisms.

Prior to the advent of gene sequencing, scientists guessed at organisms' relationships to each other by studying their physical characteristics. Organisms with similar characteristics were often assumed to be related – even though many examples were known of convergent evolution, where two unrelated organisms evolve the same traits separately.

With the advent of gene sequencing and molecular genetics- referring to the ability to read the DNA molecule at the molecular level- it became possible to trace descent lineages directly. Scientists can now read a cell's source code and determine at where, and roughly when, an organism's genome changed.

As a result, a great deal of material that was taught in schools as recently as ten years ago is now known to be incomplete. Archaea and bacteria – once classified in the same kingdom – are now known to be genetically quite different from each other. Fungi are now known to be more closely related to animals than plants. Many other fantastically weird and fascinating discoveries have come out of the genome revolution – each one bringing us a step closer to understanding what makes us who we are, and how we are interconnected.

Gene sequencing has also led to a revolution in the way we think about, diagnose, and treat disease. In many cases, it's now possible to know how likely a person is to get a given disease based on looking at their genome.

Scientists hope that this will lead to great revolutions in medicine in the centuries to come – as medicine catches up to genetics, it may someday be possible to determine what medications will work best on a disease, or what lifestyle changes will keep a person healthy, simply by reading their DNA.

This has also led to new ethical and economic challenges.

Some women whose genes have certain mutation of the BRCA1/2 gene, for example, opt to have their breasts and ovaries removed even if they are healthy – because they know there is a high chance that they will develop cancer in these organs.

In 2013, Angelina Jolie made headlines by going public with her choice to have her own breasts removed after finding out through a genetic test that she had an 87% chance of some day acquiring breast cancer.

In other cases, geneticists can tell people that they will develop a serious disease – but do not yet have the tools to stop it from happening. People in families with Huntington's disease, for example, can find out if they have the gene for this devastating and inevitably fatal dementia. But what can they do with this information?

An unexpected economic challenge has come from health insurance companies. Insurance companies have always made their money by gambling on who was likely to get sick and who wasn't. Now that the tools exist for companies to find out who is more likely to get sick at a very fine level of detail, concerns have been raised that people with unhealthy genes might be charged much more for health insurance than people with healthy genes.

Regulation of Gene Expression

Regulation of gene expression, or gene regulation, includes a wide range of mechanisms that are used by cells to increase or decrease the production of specific gene products (protein or RNA). Sophisticated programs of gene expression are widely observed in biology, for example to trigger developmental pathways, respond to environmental stimuli, or adapt to new food sources. Virtually any step of gene expression can be modulated, from transcriptional initiation, to RNA processing, and to the post-translational modification of a protein. Often, one gene regulator controls another, and so on, in a gene regulatory network.

Gene regulation is essential for viruses, prokaryotes and eukaryotes as it increases the versatility and adaptability of an organism by allowing the cell to express protein when needed. Although as early as 1951, Barbara McClintock showed interaction between two genetic loci, Activator (Ac) and Dissociator (Ds), in the color formation of maize seeds, the first discovery of a gene regulation system is widely considered to be the identification in 1961 of the lac operon, discovered by François Jacob and Jacques Monod, in which some enzymes involved in lactose metabolism are expressed by E. coli only in the presence of lactose and absence of glucose.

In multicellular organisms, gene regulation drives cellular differentiation and morphogenesis in the embryo, leading to the creation of different cell types that possess different gene expression profiles from the same genome sequence. Although this does not explain how gene regulation originated, evolutionary biologists include it as a partial explanation of how evolution works at a molecular level, and it is central to the science of evolutionary developmental biology ("evo-devo").

Regulated Stages of Gene Expression

Any step of gene expression may be modulated, from the DNA-RNA transcription step to post-translational modification of a protein. The following is a list of stages where gene expression is regulated, the most extensively utilised point is Transcription Initiation:

- Chromatin domains.

- Transcription.

- Post-transcriptional modification.

- RNA transport.

- Translation.

- mRNA degradation.

Modification of DNA

In eukaryotes, the accessibility of large regions of DNA can depend on its chromatin structure, which can be altered as a result of histone modifications directed by DNA methylation, ncRNA, or DNA-binding protein. Hence these modifications may up or down regulate the expression of a gene. Some of these modifications that regulate gene expression are inheritable and are referred to as epigenetic regulation.

Structural

Transcription of DNA is dictated by its structure. In general, the density of its packing is indicative of the frequency of transcription. Octameric protein complexes called nucleosomes are responsible for the amount of supercoiling of DNA, and these complexes can be temporarily modified by processes such as phosphorylation or more permanently modified by processes such as methylation. Such modifications are considered to be responsible for more or less permanent changes in gene expression levels.

Chemical

Methylation of DNA is a common method of gene silencing. DNA is typically methylated by methyltransferase enzymes on cytosine nucleotides in a CpG dinucleotide sequence (also called "CpG islands" when densely clustered). Analysis of the pattern of methylation in a given region of DNA (which can be a promoter) can be achieved through a method called bisulfite mapping. Methylated cytosine residues are unchanged by the treatment, whereas unmethylated ones are changed to uracil. The differences are analyzed by DNA sequencing or by methods developed to quantify SNPs, such as Pyrosequencing (Biotage) or MassArray (Sequenom), measuring the relative amounts of C/T at the CG dinucleotide. Abnormal methylation patterns are thought to be involved in oncogenesis.

Histone acetylation is also an important process in transcription. Histone acetyltransferase enzymes (HATs) such as CREB-binding protein also dissociate the DNA from the histone complex, allowing transcription to proceed. Often, DNA methylation and histone deacetylation work together in gene silencing. The combination of the two seems to be a signal for DNA to be packed more densely, lowering gene expression.

Regulation of Transcription

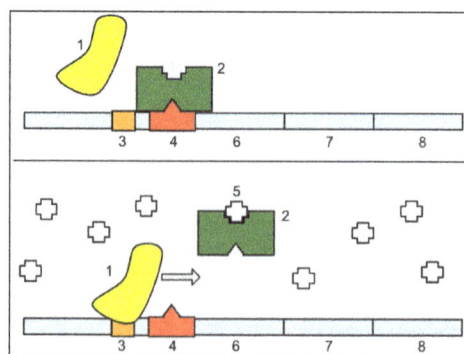

In the above figure, 1: RNA Polymerase, 2: Repressor, 3: Promoter, 4: Operator, 5: Lactose, 6: lacZ, 7: lacY, 8: lacA. Top: The gene is essentially turned off. There is no lactose to inhibit the repressor, so the repressor binds to the operator, which obstructs the RNA polymerase from binding to the promoter and making lactase. Bottom: The gene is turned on. Lactose is inhibiting the repressor, allowing the RNA polymerase to bind with the promoter, and express the genes, which synthesize lactase. Eventually, the lactase will digest all of the lactose, until there is none to bind to the repressor. The repressor will then bind to the operator, stopping the manufacture of lactase.

Regulation of transcription thus controls when transcription occurs and how much RNA is created. Transcription of a gene by RNA polymerase can be regulated by several mechanisms. Specificity factors alter the specificity of RNA polymerase for a given promoter or set of promoters, making it more or less likely to bind to them (i.e., sigma factors used in prokaryotic transcription). Repressors bind to the Operator, coding sequences on the DNA strand that are close to or overlapping the promoter region, impeding RNA polymerase's progress along the strand, thus impeding the expression of the gene.The image to the right demonstrates regulation by a repressor in the lac operon. General transcription factors position RNA polymerase at the start of a protein-coding sequence and then release the polymerase to transcribe the mRNA. Activators enhance the interaction between RNA polymerase and a particular promoter, encouraging the expression of the gene. Activators do this by increasing the attraction of RNA polymerase for the promoter, through interactions with subunits of the RNA polymerase or indirectly by changing the structure of the DNA. Enhancers are sites on the DNA helix that are bound by activators in order to loop the DNA bringing a specific promoter to the initiation complex. Enhancers are much more common in eukaryotes than prokaryotes, where only a few examples exist (to date). Silencers are regions of DNA sequences that, when bound by particular transcription factors, can silence expression of the gene.

Regulation of Transcription in Cancer

In vertebrates, the majority of gene promoters contain a CpG island with numerous CpG sites. When many of a gene's promoter CpG sites are methylated the gene becomes silenced. Colorectal cancers typically have 3 to 6 driver mutations and 33 to 66 hitchhiker or passenger mutations. However, transcriptional silencing may be of more importance than mutation in causing progression to cancer. For example, in colorectal cancers about 600 to 800 genes are transcriptionally silenced by CpG island methylation. Transcriptional repression in cancer can also occur by other epigenetic mechanisms, such as altered expression of microRNAs. In breast cancer, transcriptional repression of BRCA1 may occur more frequently by over-expressed microRNA-182 than by hypermethylation of the BRCA1 promoter.

Regulation of Transcription in Addiction

One of the cardinal features of addiction is its persistence. The persistent behavioral changes appear to be due to long-lasting changes, resulting from epigenetic alterations affecting gene expression, within particular regions of the brain. Drugs of abuse cause three types of epigenetic alteration in the brain. These are (1) histone acetylations and histone methylations, (2) DNA methylation at CpG sites, and (3) epigenetic downregulation or upregulation of microRNAs.

Chronic nicotine intake in mice alters brain cell epigenetic control of gene expression through acetylation of histones. This increases expression in the brain of the protein FosB, important in addiction. Cigarette addiction was also studied in about 16,000 humans, including never smokers, current smokers, and those who had quit smoking for up to 30 years. In blood cells, more than 18,000 CpG sites (of the roughly 450,000 analyzed CpG sites in the genome) had frequently altered methylation among current smokers. These CpG sites occurred in over 7,000 genes, or roughly a third of known human genes. The majority of the differentially methylated CpG sites returned to the level of never-smokers within five years of smoking cessation. However, 2,568 CpGs among 942 genes remained differentially methylated in former versus never smokers. Such remaining epigenetic changes can be viewed as "molecular scars" that may affect gene expression.

In rodent models, drugs of abuse, including cocaine, methampheamine, alcohol and tobacco smoke products, all cause DNA damage in the brain. During repair of DNA damages some individual repair events can alter the methylation of DNA and/or the acetylations or methylations of histones at the sites of damage, and thus can contribute to leaving an epigenetic scar on chromatin.

Such epigenetic scars likely contribute to the persistent epigenetic changes found in addiction.

Post-transcriptional Regulation

After the DNA is transcribed and mRNA is formed, there must be some sort of regulation on how much the mRNA is translated into proteins. Cells do this by modulating the capping, splicing, addition of a Poly(A) Tail, the sequence-specific nuclear export rates, and, in several contexts, sequestration of the RNA transcript. These processes occur in eukaryotes but not in prokaryotes. This modulation is a result of a protein or transcript that, in turn, is regulated and may have an affinity for certain sequences.

Three Prime Untranslated Regions and MicroRNAs

Three prime untranslated regions (3'-UTRs) of messenger RNAs (mRNAs) often contain regulatory sequences that post-transcriptionally influence gene expression. Such 3'-UTRs often contain both binding sites for microRNAs (miRNAs) as well as for regulatory proteins. By binding to specific sites within the 3'-UTR, miRNAs can decrease gene expression of various mRNAs by either inhibiting translation or directly causing degradation of the transcript. The 3'-UTR also may have silencer regions that bind repressor proteins that inhibit the expression of a mRNA.

The 3'-UTR often contains miRNA response elements (MREs). MREs are sequences to which miRNAs bind. These are prevalent motifs within 3'-UTRs. Among all regulatory motifs within the 3'-UTRs (e.g. including silencer regions), MREs make up about half of the motifs.

As of 2014, the miRBase web site, an archive of miRNA sequences and annotations, listed 28,645 entries in 233 biologic species. Of these, 1,881 miRNAs were in annotated human miRNA loci. miRNAs were predicted to have an average of about four hundred target mRNAs (affecting expression of several hundred genes). Freidman et al. estimate that >45,000 miRNA target sites within human mRNA 3'-UTRs are conserved above background levels, and >60% of human protein-coding genes have been under selective pressure to maintain pairing to miRNAs.

Direct experiments show that a single miRNA can reduce the stability of hundreds of unique mRNAs. Other experiments show that a single miRNA may repress the production of hundreds of proteins, but that this repression often is relatively mild (less than 2-fold).

The effects of miRNA dysregulation of gene expression seem to be important in cancer. For instance, in gastrointestinal cancers, a 2015 paper identified nine miRNAs as epigenetically altered and effective in down-regulating DNA repair enzymes.

The effects of miRNA dysregulation of gene expression also seem to be important in neuropsychiatric disorders, such as schizophrenia, bipolar disorder, major depressive disorder, Parkinson's disease, Alzheimer's disease and autism spectrum disorders.

Regulation of Translation

The translation of mRNA can also be controlled by a number of mechanisms, mostly at the level of initiation. Recruitment of the small ribosomal subunit can indeed be modulated by mRNA secondary structure, antisense RNA binding, or protein binding. In both prokaryotes and eukaryotes, a large number of RNA binding proteins exist, which often are directed to their target sequence by the secondary structure of the transcript, which may change depending on certain conditions, such as temperature or presence of a ligand (aptamer). Some transcripts act as ribozymes and self-regulate their expression.

Examples of Gene Regulation

- Enzyme induction is a process in which a molecule (e.g., a drug) induces (i.e., initiates or enhances) the expression of an enzyme.

- The induction of heat shock proteins in the fruit fly Drosophila melanogaster.

- The Lac operon is an interesting example of how gene expression can be regulated.

- Viruses, despite having only a few genes, possess mechanisms to regulate their gene expression, typically into an early and late phase, using collinear systems regulated by anti-terminators (lambda phage) or splicing modulators (HIV).

- Gal4 is a transcriptional activator that controls the expression of GAL1, GAL7, and GAL10 (all of which code for the metabolic of galactose in yeast). The GAL4/UAS system has been used in a variety of organisms across various phyla to study gene expression.

Developmental Biology

A large number of studied regulatory systems come from developmental biology. Examples include:

- The colinearity of the Hox gene cluster with their nested antero-posterior patterning.

- Pattern generation of the hand (digits - interdigits): the gradient of sonic hedgehog (secreted inducing factor) from the zone of polarizing activity in the limb, which creates a gradient of active Gli3, which activates Gremlin, which inhibits BMPs also secreted in the limb, results in the formation of an alternating pattern of activity as a result of this reaction-diffusion system.

- Somitogenesis is the creation of segments (somites) from a uniform tissue (Pre-somitic Mesoderm). They are formed sequentially from anterior to posterior. This is achieved in amniotes possibly by means of two opposing gradients, Retinoic acid in the anterior (wavefront) and Wnt and Fgf in the posterior, coupled to an oscillating pattern (segmentation clock) composed of FGF + Notch and Wnt in antiphase.

- Sex determination in the soma of a Drosophila requires the sensing of the ratio of autosomal genes to sex chromosome-encoded genes, which results in the production of sexless splicing factor in females, resulting in the female isoform of doublesex.

Circuitry

Up-regulation and Down-regulation

Up-regulation is a process that occurs within a cell triggered by a signal (originating internal or external to the cell), which results in increased expression of one or more genes and as a result the protein(s) encoded by those genes. Conversely, down-regulation is a process resulting in decreased gene and corresponding protein expression.

- Up-regulation occurs, for example, when a cell is deficient in some kind of receptor. In this case, more receptor protein is synthesized and transported to the membrane of the cell and, thus, the sensitivity of the cell is brought back to normal, reestablishing homeostasis.

- Down-regulation occurs, for example, when a cell is overstimulated by a neurotransmitter, hormone, or drug for a prolonged period of time, and the expression of the receptor protein is decreased in order to protect the cell.

Inducible vs. Repressible Systems

Gene Regulation can be summarized by the response of the respective system:

- Inducible systems - An inducible system is off unless there is the presence of some molecule (called an inducer) that allows for gene expression. The molecule is said to "induce expression". The manner by which this happens is dependent on the control mechanisms as well as differences between prokaryotic and eukaryotic cells.

- Repressible systems - A repressible system is on except in the presence of some molecule (called a corepressor) that suppresses gene expression. The molecule is said to "repress expression". The manner by which this happens is dependent on the control mechanisms as well as differences between prokaryotic and eukaryotic cells.

The GAL4/UAS system is an example of both an inducible and repressible system. Gal4 binds an upstream activation sequence (UAS) to activate the transcription of the GAL1/GAL7/GAL10 cassette. On the other hand, a MIG1 response to the presence of glucose can inhibit GAL4 and therefore stop the expression of the GAL1/GAL7/GAL10 cassette.

Theoretical Circuits

- Repressor/Inducer: an activation of a sensor results in the change of expression of a gene.

- Negative feedback: The gene product downregulates its own production directly or indirectly, which can result in,

  ◦ Keeping transcript levels constant/proportional to a factor.

  ◦ Inhibition of run-away reactions when coupled with a positive feedback loop.

  ◦ Creating an oscillator by taking advantage in the time delay of transcription and translation, given that the mRNA and protein half-life is shorter.

- Positive feedback: The gene product upregulates its own production directly or indirectly, which can result in,

  ◦ Signal amplification.

  ◦ Bistable switches when two genes inhibit each other and both have positive feedback.

  ◦ Pattern generation.

Study Methods

In general, most experiments investigating differential expression used whole cell extracts of RNA, called steady-state levels, to determine which genes changed and by how much. These are, however, not informative of where the regulation has occurred and may mask conflicting regulatory processes, but it is still the most commonly analysed (quantitative PCR and DNA microarray).

When studying gene expression, there are several methods to look at the various stages. In eukaryotes these include:

- The local chromatin environment of the region can be determined by ChIP-chip analysis by pulling down RNA Polymerase II, Histone 3 modifications, Trithorax-group protein, Polycomb-group protein, or any other DNA-binding element to which a good antibody is available.

- Epistatic interactions can be investigated by synthetic genetic array analysis.

- Due to post-transcriptional regulation, transcription rates and total RNA levels differ significantly. To measure the transcription rates nuclear run-on assays can be done and newer high-throughput methods are being developed, using thiol labelling instead of radioactivity.

- Only 5% of the RNA polymerised in the nucleus exits, and not only introns, abortive products, and non-sense transcripts are degradated. Therefore, the differences in nuclear and cytoplasmic levels can be see by separating the two fractions by gentle lysis.

- Alternative splicing can be analysed with a splicing array or with a tiling array.

- All in vivo RNA is complexed as RNPs. The quantity of transcripts bound to specific protein can be also analysed by RIP-Chip. For example, DCP2 will give an indication of sequestered protein; ribosome-bound gives and indication of transcripts active in transcription (although a more dated method, called polysome fractionation, is still popular in some labs)

- Protein levels can be analysed by Mass spectrometry, which can be compared only to quantitative PCR data, as microarray data is relative and not absolute.

- RNA and protein degradation rates are measured by means of transcription inhibitors (actinomycin D or α-amanitin) or translation inhibitors (Cycloheximide), respectively.

Extranuclear Inheritance

Extranuclear inheritance or cytoplasmic inheritance is the transmission of genes that occur outside the nucleus. It is found in most eukaryotes and is commonly known to occur in cytoplasmic organelles such as mitochondria and chloroplasts or from cellular parasites like viruses or bacteria.

Organelles

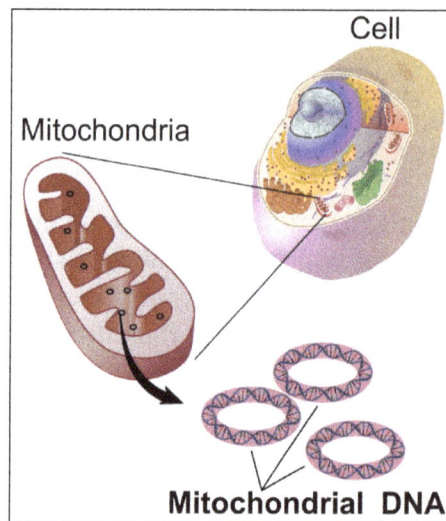

Mitochondria contain their own DNA.

Mitochondria are organelles which function to transform energy as a result of cellular respiration. Chloroplasts are organelles which function to produce sugars via photosynthesis in plants and algae. The genes located in mitochondria and chloroplasts are very important for proper cellular function, yet the genomes replicate independently of the DNA located in the nucleus, which is typically arranged in chromosomes that only replicate one time preceding cellular division. The extranuclear genomes of mitochondria and chloroplasts however replicate independently of cell division. They replicate in response to a cell's increasing energy needs which adjust during that cell's lifespan. Since they replicate independently, genomic recombination of these genomes is rarely found in offspring, contrary to nuclear genomes in which recombination is common. Mitochondrial diseases are inherited from the mother, not from the father: mitochondria with their mitochondrial DNA from the mother's egg cell are incorporated into the zygote and passed to daughter cells, whereas those from the sperm are not.

Parasites

Extranuclear transmission of viral genomes and symbiotic bacteria is also possible. An example of viral genome transmission is perinatal transmission. This occurs from mother to fetus during the perinatal period, which begins before birth and ends about 1 month after birth. During this time viral material may be passed from mother to child in the bloodstream or breastmilk. This is of particular concern with mothers carrying HIV or Hepatitis C viruses. Symbiotic cytoplasmic bacteria are also inherited in organisms such as insects and protists.

Types

Three general types of extranuclear inheritance exist.

- Vegetative segregation results from random replication and partitioning of cytoplasmic organelles. It occurs with chloroplasts and mitochondria during mitotic cell divisions and results in daughter cells that contain a random sample of the parent cell's organelles. An example of vegetative segregation is with mitochondria of asexually replicating yeast cells.

- Uniparental inheritance occurs in extranuclear genes when only one parent contributes organellar DNA to the offspring. A classic example of uniparental gene transmission is the maternal inheritance of human mitochondria. The mother's mitochondria are transmitted to the offspring at fertilization via the egg. The father's mitochondrial genes are not transmitted to the offspring via the sperm. Very rare cases which require further investigation have been reported of paternal mitochondrial inheritance in humans, in which the father's mitochondrial genome is found in offspring. Chloroplast genes can also inherit uniparentally during sexual reproduction. They are historically thought to inherit maternally, but paternal inheritance in many species is increasingly being identified. The mechanisms of uniparental inheritance from species to species differ greatly and are quite complicated. For instance, chloroplasts have been found to exhibit maternal, paternal and biparental modes even within the same species.

- Biparental inheritance occurs in extranuclear genes when both parents contribute organellar DNA to the offspring. It may be less common than uniparental extranuclear inheritance, and usually occurs in a permissible species only a fraction of the time. An example of biparental mitochondrial inheritance is in the yeast *Saccharomyces cerevisiae*. When two haploid cells of opposite mating type fuse they can both contribute mitochondria to the resulting diploid offspring.

Mutant Mitochondria

Poky is a mutant of the fungus Neurospora crassa that has extranuclear inheritance. Poky is characterized by slow growth, a defect in mitochondrial ribosome assembly and deficiencies in several cytochromes. The studies of poky mutants were among the first to establish an extranuclear mitochondrial basis for inheritance of a particular genotype. It was initially found, using genetic crosses, that poky is maternally inherited. Subsequently, the primary defect in the poky mutants was determined to be a deletion in the mitochondrial DNA sequence encoding the small subunit of mitochondrial ribosomal RNA.

Gene Transfer Agent

Gene transfer agents (GTAs) are DNA-containing virus-like particles that are produced by some bacteria and archaea and mediate horizontal gene transfer. Different GTA types have originated independently from viruses in several bacterial and archaeal lineages. These cells produce GTA particles containing short segments of the DNA present in the cell. After the particles are released from the producer cell, they can attach to related cells and inject their DNA into the cytoplasm. The DNA can then become part of the recipient cells' genome.

Discovery of Gene Transfer Agents

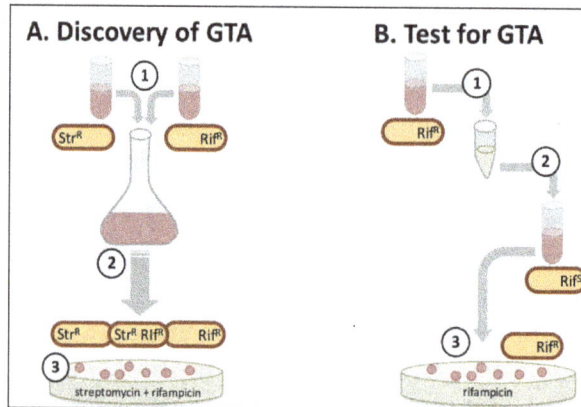

Methods for detecting gene transfer agents. (A) Method used by Marrs in 1974. (B) Cell-free extract method.

The first GTA system was discovered in 1974, when mixed cultures of *Rhodobacter capsulatus* strains produced a high frequency of cells with new combinations of genes. The factor responsible was distinct from known gene-transfer mechanisms in being independent of cell contact, insensitive to deoxynuclease, and not associated with phage production. Because of its presumed function it was named gene transfer agent (GTA, now RcGTA) More recently other gene transfer agent systems have been discovered by incubating filtered (cell-free) culture medium with a genetically distinct strain.

GTA Genes and Evolution

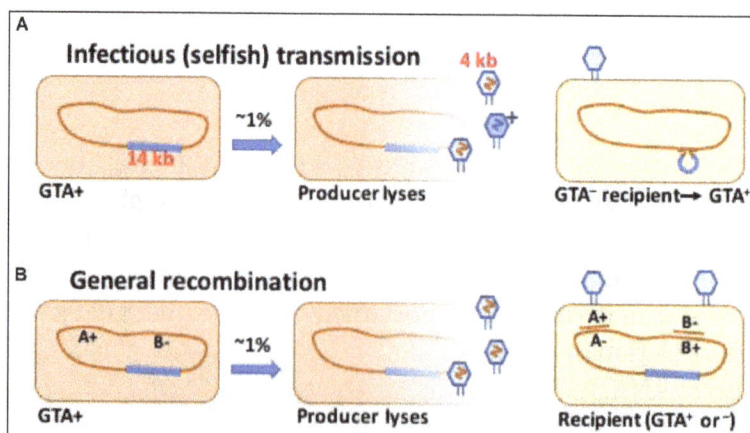

GTA gene clusters.

The genes specifying GTAs are derived from bacteriophage (phage) DNA that has integrated into a host chromosome. Such prophages often acquire mutations that make them defective and unable to produce phage particles. Many bacterial genomes contain one or more defective prophages that have undergone more-or less-extensive mutation and deletion. Gene transfer agents, like defective prophages, arise by mutation of prophages, but they retain functional genes for the head and tail components of the phage particle (structural genes) and the genes for DNA packaging. The phage genes specifying its regulation and DNA replication have typically been deleted, and expression of the cluster of structural genes is under the control of cellular regulatory systems. Additional genes that contribute to GTA production or uptake are usually present at other chromosome locations. Some of these have regulatory functions, and others contribute directly to GTA production (*e.g.* the phage-derived lysis genes) or uptake and recombination (*e.g.* production of cell-surface capsule and DNA transport proteins) These GTA-associated genes are often under coordinated regulation with the main GTA gene cluster. Phage-derived cell-lysis proteins (holin and endolysin) then weaken the cell wall and membrane, allowing the cell to burst and release the GTA particles. The number of GTA particles produced by each cell is not known.

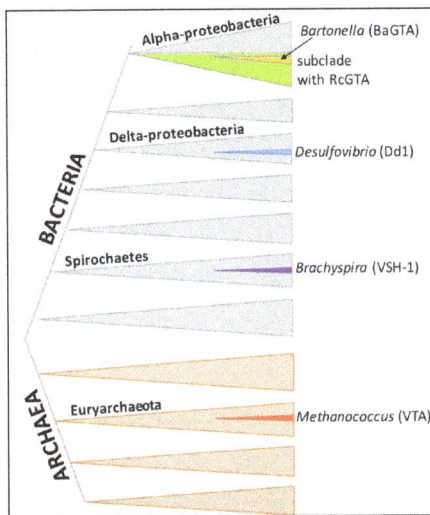

Schematic diagram of phylogenetic relationships between known bacterial gene transfer agents.

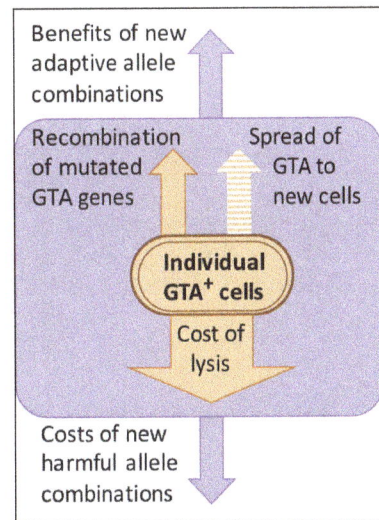

The evolutionary forces that act on bacterial gene transfer agent and the cells that produce it.

Some GTA systems appear to be recent additions to their host genomes, but others have been maintained for many millions of years. Where studies of sequence divergence have been done (dN/dS analysis), they indicate that the genes are being maintained by natural selection for protein function (i.e. defective versions are being eliminated).

However the nature of this selection is not clear. Although the discoverers of GTA assumed that gene transfer was the function of the particles, the presumed benefits of gene transfer come at a substantial cost to the population. Most of this cost arises because GTA-producing cells must lyse (burst open) to release their GTA particles, but there are also genetic costs associated with making new combinations of genes, because most new combinations will usually be less fit than the original combination. One alternative explanation is that GTA genes persist because GTAs are genetic parasites that spread infectiously to new cells. However this is ruled out because GTA particles typically are too small to contain the genes that encode them. For example, the main RcGTA cluster is 14 kb long, but RcGTA particles can contain only 4–5 kb of DNA.

Most bacteria have not been screened for the presence of GTAs, and many more GTA systems may await discovery. Although DNA-based surveys for GTA-related genes have found homologs in many genomes, but interpretation is hindered by the difficulty of distinguishing genes that encode GTAs from ordinary prophage genes.

GTA Production

In laboratory cultures, production of GTAs is typically maximized by particular growth conditions that induce transcription of the GTA genes; most GTAs are not induced by the DNA-damaging treatments that induce many prophages. Even under maximally inducing conditions only a small fraction of the culture produces GTAs, typically less than 1%.

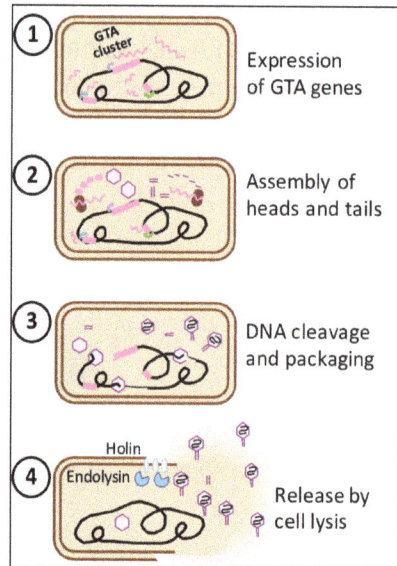

Typical steps in the production of bacterial gene transfer agents. (1) Transcription and translation of the GTA genes. (2) Assembly of GTA structural proteins into empty heads and unattached tails. (3) Packaging of 'headful' segments of DNA into heads and attachment of tails. (4) Lysis of the cell.

The steps in GTA production are derived from those of phage infection. The structural genes are first transcribed and translated, and the proteins assembled into empty heads and unattached tails. The DNA packaging machinery then packs DNA into each head, cutting the DNA when the head is full, attaching a tail to the head, and then moving the newly-created DNA end on to a new empty head. Unlike prophage genes, the genes encoding GTAs are not excised from the genome and replicated for packaging in GTA particles. The two best studied GTAs (RcGTA and BaGTA) randomly package all of the DNA in the cell, with no overrepresentation of GTA-encoding genes. The number of GTA particles produced by each cell is not known.

GTA-mediated Transduction

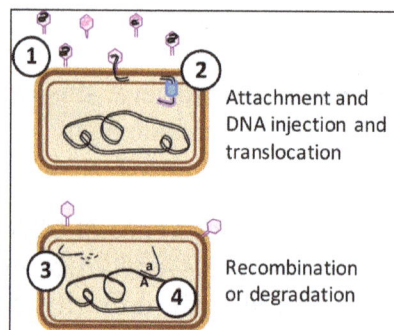

Genetic transduction by bacterial gene transfer agents. (1) GTA particles encounter a suitable recipient cell. (2) Particles attach to cell and inject their DNA, and cellular proteins translocate the DNA across the inner membrane. (3) DNA is degraded if it cannot recombine with the recipient genome. (4) DNA with similar sequences to the recipient genome undergoes recombination.

Whether release of GTA particles leads to transfer of DNA to new genomes depends on several factors. First, the particles must survive in the environment – little is known about this, although particles are reported to be quite unstable under laboratory conditions. Second, particles must encounter and attach to suitable recipient cells, usually members of the same or a closely related species. Like phages, GTAs attach to specific protein or carbohydrate structures on the recipient cell surface before injecting their DNA. Unlike phage, the well-studied GTAs appear to inject their DNA only across the first of the two membranes surrounding the recipient cytoplasm, and they use a different system, competence-derived rather than phage-derived, to transport one strand of the double-stranded DNA across the inner membrane into the cytoplasm.

If the cell's recombinational repair machinery finds a chromosomal sequence very similar to the incoming DNA, it replaces the former with the latter by homologous recombination, mediated by the cell's RecA protein. I the sequences are not identical this will produce a cell with a new genetic combination. However, if the incoming DNA is not closely related to DNA sequences in the cell it will be degraded, and the cell will reuse its nucleotides for DNA replication.

Specific GTA Systems

RcGTA (Rhodobacter Capsulatus)

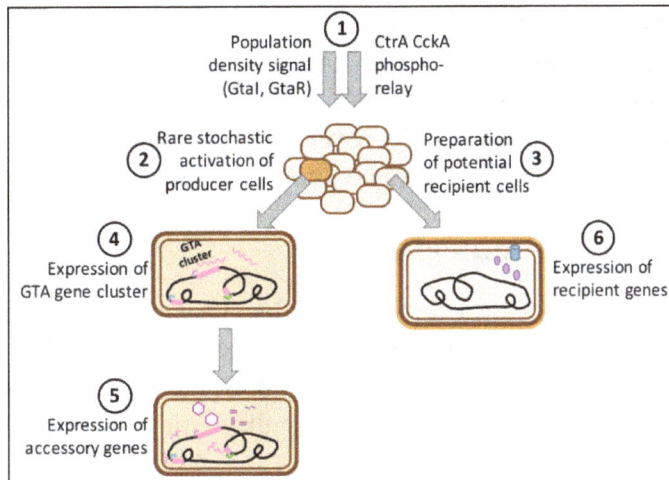

Regulation diagram for RcGTA, the Rhodobacter capsulatus gene transfer agent.

The GTA produced by the alphaproteobacterium *Rhodobacter capsulatus*, named *R. capsulatus* GTA (RcGTA), is currently the best studied GTA. When laboratory cultures of *R. capsulatus* enter stationary phase, a subset of the bacterial population induces production of RcGTA, and the particles are subsequently released from the cells through cell lysis. Most of the RcGTA structural genes are encoded in a ~ 15 kb genetic cluster on the bacterial chromosome. However, other genes required for RcGTA function, such as the genes required for cell lysis, are located separately. RcGTA particles contain 4.5 kb DNA fragments, with even representation of the whole chromosome except for a 2-fold dip at the site of the RcGTA gene cluster.

Regulation of GTA production and transduction has been best studied in *R. capsulatus*, where a quorum-sensing system and a CtrA-phosphorelay control expression of not only the main RcGTA gene cluster, but also a holin/endolysin cell lysis system, particle head spikes, an attachment protein (possibly tail fibers), and the capsule and DNA processing genes needed for RcGTA recipient function. An uncharacterized stochastic process further limits expression of the gene cluster is to only 0.1-3% of the cells.

RcGTA-like clusters are found in a large subclade of the alphaproteobacteria, although the genes also appear to be frequently lost by deletion. Recently, several members of the order Rhodobacterales have been demonstrated to produce functional RcGTA-like particles. Groups of genes with homology to the RcGTA are present in the chromosomes of various types of alphaproteobacteria.

DsGTA (Dinoroseobacter Shibae)

D. shibae, like R. capsulatus, is a member of the Order Rhodobacterales, and its GTA shares a common ancestor and many features with RcGTA, including gene organization, packaging of short DNA fragments (4.2 kb) and regulation by quorum sensing and a CtrA phosphorelay. However, its DNA packaging machinery has much more specificity, with sharp peaks and valleys of coverage suggestion that it may preferentially initiate packaging at specific sites in the genome. The DNA of the major DsGTA gene cluster is packaged very poorly.

BaGTA (Bartonella Species)

Bartonella species are members of the Alphaproteobacteria like R. capsulatus and D. shibae, but BaGTA is not related to RcGTA and DsGTA. BaGTA particles are larger than RcGTA and contain 14 kb DNA fragments. Although this capacity could in principle allow BaGTA to package and transmit its 14 kb GTA cluster, measurements of DNA coverage show reduced coverage of the cluster. An adjacent region of high coverage is thought to be due to local DNA replication.

VSH-1 (Brachyspira Hyodysenteriae)

Brachyspira is a genus of spirochete; several species have been shown to carry homologous GTA gene clusters. Particles contain 7.5 kb DNA fragments. Production of VSH-1 is stimulated by the DNA-damaging agent mitomycin C and by some antibiotics. It is also associated with detectable cell lysis, indicating that a substantial fraction of the culture may be producing VSH-1.

Dd1 (Desulfovibriondesulfuricans)

D. desulfuricans is a soil bacterium in the deltaproteobacteria; Dd1 packages 13.6 kb DNA fragments

VTA (Methanococcus Voltae)

M. voltae is an archaean; its GTA is known to transfer 4.4 kb DNA fragments but has not been otherwise characterized.

Genomics

Genomics is an interdisciplinary field of biology focusing on the structure, function, evolution, mapping, and editing of genomes. A genome is an organism's complete set of DNA, including all of its genes. In contrast to genetics, which refers to the study of individual genes and their roles in inheritance, genomics aims at the collective characterization and quantification of all of an organism's genes, their interrelations and influence on the organism. Genes may direct the production of proteins with the assistance of enzymes and messenger molecules. In turn, proteins make up body structures such as organs and tissues as well as control chemical reactions and carry signals between cells. Genomics also involves the sequencing and analysis of genomes through uses of high throughput DNA sequencing and bioinformatics to assemble and analyze the function and structure of entire genomes. Advances in genomics have triggered a revolution in discovery-based research and systems biology to facilitate understanding of even the most complex biological systems such as the brain.

The field also includes studies of intragenomic (within the genome) phenomena such as epistasis (effect of one gene on another), pleiotropy (one gene affecting more than one trait), heterosis (hybrid vigour), and other interactions between loci and alleles within the genome.

Genome Analysis

After an organism has been selected, genome projects involve three components: the sequencing of DNA, the assembly of that sequence to create a representation of the original chromosome, and the annotation and analysis of that representation.

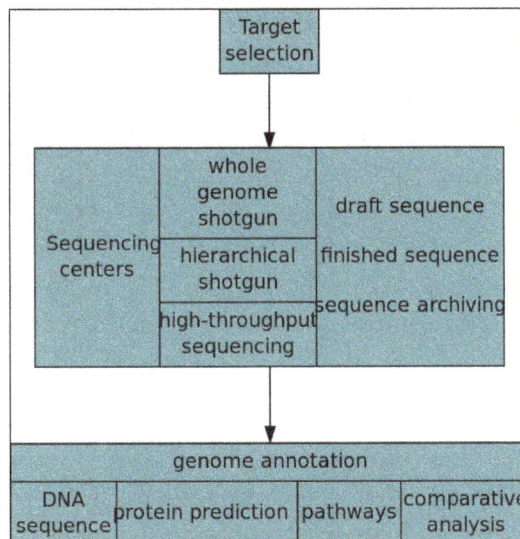

Overview of a genome project.

First, the genome must be selected, which involves several factors including cost and relevance. Second, the sequence is generated and assembled at a given sequencing center (such as BGI or DOE JGI). Third, the genome sequence is annotated at several levels: DNA, protein, gene pathways, or comparatively.

Sequencing

Historically, sequencing was done in sequencing centers, centralized facilities (ranging from large independent institutions such as Joint Genome Institute which sequence dozens of terabases a year, to local molecular biology core facilities) which contain research laboratories with the costly instrumentation and technical support necessary. As sequencing technology continues to improve, however, a new generation of effective fast turnaround benchtop sequencers has come within reach of the average academic laboratory. On the whole, genome sequencing approaches fall into two broad categories, shotgun and high-throughput (or next-generation) sequencing.

Shotgun Sequencing

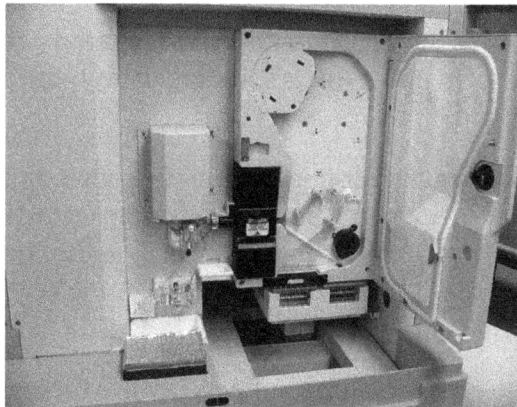

An ABI PRISM 3100 Genetic Analyzer. Such capillary sequencers automated early large-scale genome sequencing efforts.

Shotgun sequencing is a sequencing method designed for analysis of DNA sequences longer than 1000 base pairs, up to and including entire chromosomes. It is named by analogy with the rapidly expanding, quasi-random firing pattern of a shotgun. Since gel electrophoresis sequencing can only be used for fairly short sequences (100 to 1000 base pairs), longer DNA sequences must be broken into random small segments which are then sequenced to obtain *reads*. Multiple overlapping reads for the target DNA are obtained by performing several rounds of this fragmentation and sequencing. Computer programs then use the overlapping ends of different reads to assemble them into a continuous sequence. Shotgun sequencing is a random sampling process, requiring over-sampling to ensure a given nucleotide is represented in the reconstructed sequence; the average number of reads by which a genome is over-sampled is referred to as coverage.

For much of its history, the technology underlying shotgun sequencing was the classical chain-termination method or 'Sanger method', which is based on the selective incorporation of chain-terminating dideoxynucleotides by DNA polymerase during in vitro DNA replication. Recently, shotgun sequencing has been supplanted by high-throughput sequencing methods, especially for large-scale, automated genome analyses. However, the Sanger method remains in wide use, primarily for smaller-scale projects and for obtaining especially long contiguous DNA sequence reads (>500 nucleotides). Chain-termination methods require a single-stranded DNA template, a DNA primer, a DNA polymerase, normal deoxynucleosidetriphosphates (dNTPs), and modified nucleotides (dideoxyNTPs) that terminate DNA strand elongation. These chain-terminating nucleotides lack a 3'-OH group required for the formation of a phosphodiester bond between two nucleotides, causing DNA polymerase to cease extension of DNA when a ddNTP is incorporated. The ddNTPs may be radioactively or fluorescently labelled for detection in DNA sequencers. Typically, these machines can sequence up to 96 DNA samples in a single batch (run) in up to 48 runs a day.

High-throughput Sequencing

The high demand for low-cost sequencing has driven the development of high-throughput sequencing technologies that parallelize the sequencing process, producing thousands or millions of sequences at once. High-throughput sequencing is intended to lower the cost of DNA sequencing beyond what is possible with standard dye-terminator methods. In ultra-high-throughput sequencing, as many as 500,000 sequencing-by-synthesis operations may be run in parallel.

Illumina Genome Analyzer II System. Illumina technologies
have set the standard for high-throughput massively parallel sequencing.

The Illumina dye sequencing method is based on reversible dye-terminators and was developed in 1996 at the Geneva Biomedical Research Institute, by Pascal Mayer and Laurent Farinelli. In this method, DNA molecules and primers are first attached on a slide and amplified with polymerase so that local clonal colonies, initially coined "DNA colonies", are formed. To determine the sequence, four types of reversible terminator bases (RT-bases) are added and non-incorporated nucleotides are washed away. Unlike pyrosequencing, the DNA chains are extended one nucleotide at a time and image acquisition can be performed at a delayed moment, allowing for very large arrays of DNA colonies to be captured by sequential images taken from a single camera. Decoupling the enzymatic reaction and the image capture allows for optimal throughput and theoretically unlimited sequencing capacity; with an optimal configuration, the ultimate throughput of the instrument depends only on the A/D conversion rate of the camera. The camera takes images of the fluorescently labeled nucleotides, then the dye along with the terminal 3' blocker is chemically removed from the DNA, allowing the next cycle.

An alternative approach, ion semiconductor sequencing, is based on standard DNA replication chemistry. This technology measures the release of a hydrogen ion each time a base is incorporated. A microwell containing template DNA is flooded with a single nucleotide, if the nucleotide is complementary to the template strand it will be incorporated and a hydrogen ion will be released. This release triggers an ISFET ion sensor. If a homopolymer is present in the template sequence multiple nucleotides will be incorporated in a single flood cycle, and the detected electrical signal will be proportionally higher.

Assembly

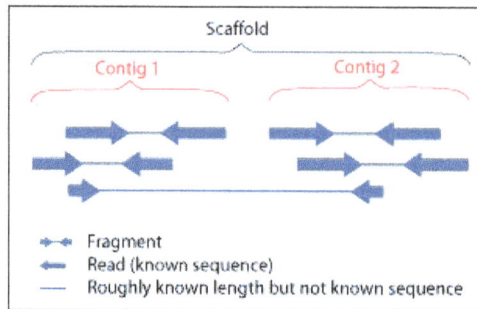

Overlapping reads form contigs; contigs and
gaps of known length form scaffolds.

Sequence assembly refers to aligning and merging fragments of a much longer DNA sequence
in order to reconstruct the original sequence. This is needed as current DNA sequencing technology cannot read whole genomes as a continuous sequence, but rather reads small pieces of
between 20 and 1000 bases, depending on the technology used. Third generation sequencing
technologies such as PacBio or Oxford Nanopore routinely generate sequenceing reads >10 kb
in length; however, they have a high error rate at approximately 15 percent. Typically the short
fragments, called reads, result from shotgun sequencing genomic DNA, or gene transcripts
(ESTs).

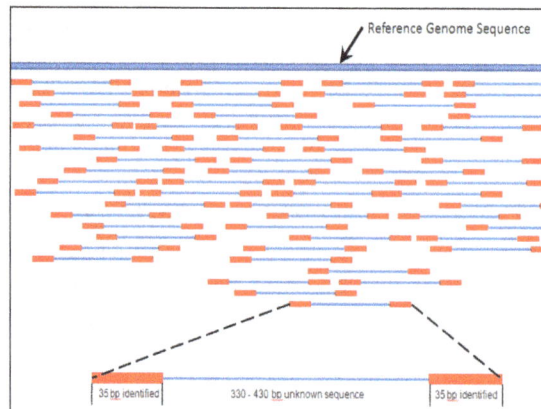

Paired end reads of next generation sequencing
data mapped to a reference genome.

Multiple, fragmented sequence reads must be assembled together on the basis of their overlapping
areas.

Assembly Approaches

Assembly can be broadly categorized into two approaches: *de novo* assembly, for genomes which
are not similar to any sequenced in the past, and comparative assembly, which uses the existing sequence of a closely related organism as a reference during assembly. Relative to comparative assembly, *de novo* assembly is computationally difficult (NP-hard), making it less favourable
for short-read NGS technologies. Within the *de novo* assembly paradigm there are two primary
strategies for assembly, Eulerian path strategies, and overlap-layout-consensus (OLC) strategies.
OLC strategies ultimately try to create a Hamiltonian path through an overlap graph which is an

NP-hard problem. Eulerian path strategies are computationally more tractable because they try to find a Eulerian path through a deBruijn graph.

Finishing

Finished genomes are defined as having a single contiguous sequence with no ambiguities representing each replicon.

Annotation

The DNA sequence assembly alone is of little value without additional analysis. Genome annotation is the process of attaching biological information to sequences, and consists of three main steps:

1. Identifying portions of the genome that do not code for proteins

2. Identifying elements on the genome, a process called gene prediction, and

3. Attaching biological information to these elements.

Automatic annotation tools try to perform these steps *in silico*, as opposed to manual annotation (curation) which involves human expertise and potential experimental verification. Ideally, these approaches co-exist and complement each other in the same annotation pipeline.

Traditionally, the basic level of annotation is using BLAST for finding similarities, and then annotating genomes based on homologues. More recently, additional information is added to the annotation platform. The additional information allows manual annotators to deconvolute discrepancies between genes that are given the same annotation. Some databases use genome context information, similarity scores, experimental data, and integrations of other resources to provide genome annotations through their Subsystems approach. Other databases (e.g. Ensembl) rely on both curated data sources as well as a range of software tools in their automated genome annotation pipeline. Structural annotation consists of the identification of genomic elements, primarily ORFs and their localisation, or gene structure. Functional annotation consists of attaching biological information to genomic elements.

Sequencing Pipelines and Databases

The need for reproducibility and efficient management of the large amount of data associated with genome projects mean that computational pipelines have important applications in genomics.

Applications of Genomics

Genomics has provided applications in many fields, including medicine, biotechnology, anthropology and other social sciences.

Genomic Medicine

Next-generation genomic technologies allow clinicians and biomedical researchers to drastically increase the amount of genomic data collected on large study populations. When combined with new informatics approaches that integrate many kinds of data with genomic data in disease research, this allows researchers to better understand the genetic bases of drug response and disease. For example, the All of Us research program aims to collect genome sequence data from 1 million participants to become a critical component of the precision medicine research platform.

Synthetic Biology and Bioengineering

The growth of genomic knowledge has enabled increasingly sophisticated applications of synthetic biology. In 2010 researchers at the J. Craig Venter Institute announced the creation of a partially synthetic species of bacterium, Mycoplasma laboratorium, derived from the genome of Mycoplasma genitalium.

Conservation Genomics

Conservationists can use the information gathered by genomic sequencing in order to better evaluate genetic factors key to species conservation, such as the genetic diversity of a population or whether an individual is heterozygous for a recessive inherited genetic disorder. By using genomic data to evaluate the effects of evolutionary processes and to detect patterns in variation throughout a given population, conservationists can formulate plans to aid a given species without as many variables left unknown as those unaddressed by standard genetic approaches.

Genetic Linkage

Genetic linkage is the tendency of DNA sequences that are close together on a chromosome to be inherited together during the meiosis phase of sexual reproduction. Two genetic markers that are physically near to each other are unlikely to be separated onto different chromatids during chromosomal crossover, and are therefore said to be more *linked* than markers that are far apart. In other words, the nearer two genes are on a chromosome, the lower the chance of recombination between them, and the more likely they are to be inherited together. Markers on different chromosomes are perfectly *unlinked*.

Genetic linkage is the most prominent exception to Gregor Mendel's Law of Independent Assortment. The first experiment to demonstrate linkage was carried out in 1905. At the time, the reason why certain traits tend to be inherited together was unknown. Later work revealed that genes are physical structures related by physical distance.

The typical unit of genetic linkage is the centimorgan (cM). A distance of 1 cM between two markers means that the markers are separated to different chromosomes on average once per 100 meiotic product, thus once per 50 meioses.

Discovery

Gregor Mendel's Law of Independent Assortment states that every trait is inherited independently of every other trait. But shortly after Mendel's work was rediscovered, exceptions to this rule were found. In 1905, the British geneticists William Bateson, Edith Rebecca Saunders and Reginald Punnett cross-bred pea plants in experiments similar to Mendel's. They were interested in trait inheritance in the sweet pea and were studying two genes—the gene for flower colour (*P*, purple, and *p*, red) and the gene affecting the shape of pollen grains (*L*, long, and *l*, round). They crossed the pure lines *PPLL* and *ppll* and then self-crossed the resulting *PpLl* lines.

According to Mendelian genetics, the expected phenotypes would occur in a 9:3:3:1 ratio of PL:Pl:pL:pl. To their surprise, they observed an increased frequency of PL and pl and a decreased frequency of Pl and pL:

Bateson, Saunders, and Punnett experiment		
Phenotype and genotype	Observed	Expected from 9:3:3:1 ratio
Purple, long (P_L_)	284	216
Purple, round (P_ll)	21	72
Red, long (ppL_)	21	72
Red, round (ppll)	55	24

Their experiment revealed linkage between the *P* and *L* alleles and the *p* and *l* alleles. The frequency of *P* occurring together with *L* and *p* occurring together with *l* is greater than that of the recombinant *Pl* and *pL*. The recombination frequency is more difficult to compute in an F2 cross than a backcross, but the lack of fit between observed and expected numbers of progeny in the above table indicate it is less than 50%. This indicated that two factors interacted in some way to create this difference by masking the appearance of the other two phenotypes. This led to the conclusion that some traits are related to each other because of their near proximity to each other on a chromosome. This provided the grounds to determine the difference between independent and codependent alleles.

The understanding of linkage was expanded by the work of Thomas Hunt Morgan. Morgan's observation that the amount of crossing over between linked genes differs led to the idea that crossover frequency might indicate the distance separating genes on the chromosome. The centimorgan, which expresses the frequency of crossing over, is named in his honour.

Linkage Map

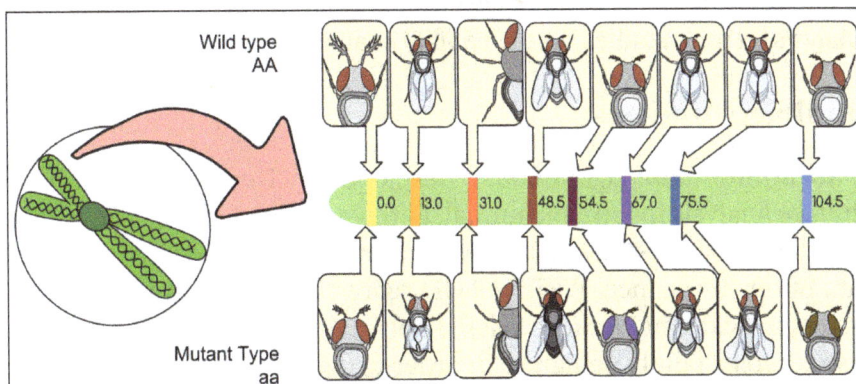

Thomas Hunt Morgan's Drosophila melanogaster genetic linkage map.

This was the first successful gene mapping work and provides important evidence for the chromosome theory of inheritance. The map shows the relative positions of alleles on the second Drosophila chromosome. The distances between the genes (centimorgans) are equal to the percentages of chromosomal crossover events that occur between different alleles.

A linkage map (also known as a genetic map) is a table for a species or experimental population that shows the position of its known genes or genetic markers relative to each other in terms of recombination frequency, rather than a specific physical distance along each chromosome. Linkage maps were first developed by Alfred Sturtevant, a student of Thomas Hunt Morgan.

A linkage map is a map based on the frequencies of recombination between markers during crossover of homologous chromosomes. The greater the frequency of recombination (segregation) between two genetic markers, the further apart they are assumed to be. Conversely, the lower the frequency of recombination between the markers, the smaller the physical distance between them. Historically, the markers originally used were detectable phenotypes (enzyme production, eye colour) derived from coding DNA sequences; eventually, confirmed or assumed noncoding DNA sequences such as microsatellites or those generating restriction fragment length polymorphisms (RFLPs) have been used.

Linkage maps help researchers to locate other markers, such as other genes by testing for genetic linkage of the already known markers. In the early stages of developing a linkage map, the data are used to assemble linkage groups, a set of genes which are known to be linked. As knowledge advances, more markers can be added to a group, until the group covers an entire chromosome. For well-studied organisms the linkage groups correspond one-to-one with the chromosomes.

A linkage map is not a physical map (such as a radiation reduced hybrid map) or gene map.

Linkage Analysis

Linkage analysis is a genetic method that searches for chromosomal segments that cosegregate with the ailment phenotype through families and is the analysis technique that has been used to determine the bulk of lipodystrophy genes. It can be used to map genes for both binary and quantitative traits. Linkage analysis may be either parametric (if we know the relationship between phenotypic and genetic similarity) or non-parametric. Parametric linkage analysis is the traditional approach, whereby the probability that a gene important for a disease is linked to a genetic marker is studied through the LOD score, which assesses the probability that a given pedigree, where the disease and the marker are cosegregating, is due to the existence of linkage (with a given linkage value) or to chance. Non-parametric linkage analysis, in turn, studies the probability of an allele being identical by descent with itself.

Parametric Linkage Analysis

The LOD score (logarithm (base 10) of odds), developed by Newton Morton, is a statistical test often used for linkage analysis in human, animal, and plant populations. The LOD score compares the likelihood of obtaining the test data if the two loci are indeed linked, to the likelihood of observing the same data purely by chance. Positive LOD scores favour the presence of linkage, whereas negative LOD scores indicate that linkage is less likely. Computerised LOD score analysis is a simple way to analyse complex family pedigrees in order to determine the linkage between Mendelian traits (or between a trait and a marker, or two markers).

The method is described in greater detail by Strachan and Read. Briefly, it works as follows:

1. Establish a pedigree.
2. Make a number of estimates of recombination frequency.

3. Calculate a LOD score for each estimate.

4. The estimate with the highest LOD score will be considered the best estimate.

The LOD score is calculated as follows:

$$LOD = Z = \log_{10} \frac{\text{probability of birth sequence with a given linkage value}}{\text{probability of birth sequence with no linkage}} = \log_{10} \frac{(1-\theta)^{NR} \times \theta^{R}}{0.5^{(NR+R)}}$$

NR denotes the number of non-recombinant offspring, and R denotes the number of recombinant offspring. The reason 0.5 is used in the denominator is that any alleles that are completely unlinked (e.g. alleles on separate chromosomes) have a 50% chance of recombination, due to independent assortment. 'θ' is the recombinant fraction, i.e. the fraction of births in which recombination has happened between the studied genetic marker and the putative gene associated with the disease. Thus, it is equal to R / (NR + R).

By convention, a LOD score greater than 3.0 is considered evidence for linkage, as it indicates 1000 to 1 odds that the linkage being observed did not occur by chance. On the other hand, a LOD score less than -2.0 is considered evidence to exclude linkage. Although it is very unlikely that a LOD score of 3 would be obtained from a single pedigree, the mathematical properties of the test allow data from a number of pedigrees to be combined by summing their LOD scores. A LOD score of 3 translates to a p-value of approximately 0.05, and no multiple testing correction (e.g. Bonferroni correction) is required.

Limitations

Linkage analysis has a number of methodological and theoretical limitations that can significantly increase the type-1 error rate and reduce the power to map human quantitative trait loci (QTL). While linkage analysis was successfully used to identify genetic variants that contribute to rare disorders such as Huntington disease, it didn't perform that well when applied to more common disorders such as heart disease or different forms of cancer. An explanation for this is that the genetic mechanisms affecting common disorders are different from those causing rare disorders.

Recombination Frequency

Recombination frequency is a measure of genetic linkage and is used in the creation of a genetic linkage map. Recombination frequency (θ) is the frequency with which a single chromosomal crossover will take place between two genes during meiosis. A centimorgan (cM) is a unit that describes a recombination frequency of 1%. In this way we can measure the genetic distance between two loci, based upon their recombination frequency. This is a good estimate of the real distance. Double crossovers would turn into no recombination. In this case we cannot tell if crossovers took place. If the loci we're analysing are very close (less than 7 cM) a double crossover is very unlikely. When distances become higher, the likelihood of a double crossover increases. As the likelihood of a double crossover increases we systematically underestimate the genetic distance between two loci.

During meiosis, chromosomes assort randomly into gametes, such that the segregation of alleles of one gene is independent of alleles of another gene. This is stated in Mendel's Second Law and is known as the law of independent assortment. The law of independent assortment always holds true for genes that are located on different chromosomes, but for genes that are on the same chromosome, it does not always hold true.

As an example of independent assortment, consider the crossing of the pure-bred homozygote parental strain with genotype *AABB* with a different pure-bred strain with genotype *aabb*. A and a and B and b represent the alleles of genes A and B. Crossing these homozygous parental strains will result in F1 generation offspring that are double heterozygotes with genotype AaBb. The F1 offspring AaBb produces gametes that are *AB*, *Ab*, *aB*, and *ab* with equal frequencies (25%) because the alleles of gene A assort independently of the alleles for gene B during meiosis. Note that 2 of the 4 gametes (50%)—*Ab* and *aB*—were not present in the parental generation. These gametes represent recombinant gametes. Recombinant gametes are those gametes that differ from both of the haploid gametes that made up the original diploid cell. In this example, the recombination frequency is 50% since 2 of the 4 gametes were recombinant gametes.

The recombination frequency will be 50% when two genes are located on different chromosomes or when they are widely separated on the same chromosome. This is a consequence of independent assortment.

When two genes are close together on the same chromosome, they do not assort independently and are said to be linked. Whereas genes located on different chromosomes assort independently and have a recombination frequency of 50%, linked genes have a recombination frequency that is less than 50%.

As an example of linkage, consider the classic experiment by William Bateson and Reginald Punnett. They were interested in trait inheritance in the sweet pea and were studying two genes—the gene for flower colour (*P*, purple, and *p*, red) and the gene affecting the shape of pollen grains (*L*, long, and *l*, round). They crossed the pure lines *PPLL* and *ppll* and then self-crossed the resulting *PpLl* lines. According to Mendelian genetics, the expected phenotypes would occur in a 9:3:3:1 ratio of PL:Pl:pL:pl. To their surprise, they observed an increased frequency of PL and pl and a decreased frequency of Pl and pL.

Bateson and Punnett experiment		
Phenotype and genotype	Observed	Expected from 9:3:3:1 ratio
Purple, long (P_L_)	284	216
Purple, round (P_ll)	21	72
Red, long (ppL_)	21	72
Red, round (ppll)	55	24

Their experiment revealed linkage between the *P* and *L* alleles and the *p* and *l* alleles. The frequency of *P* occurring together with *L* and with *p* occurring together with *l* is greater than that of the recombinant *Pl* and *pL*. The recombination frequency is more difficult to compute in an F2 cross than a backcross, but the lack of fit between observed and expected numbers of progeny in the above table indicate it is less than 50%.

The progeny in this case received two dominant alleles linked on one chromosome (referred to as coupling or cis arrangement). However, after crossover, some progeny could have received one parental chromosome with a dominant allele for one trait (e.g. Purple) linked to a recessive allele for a second trait (e.g. round) with the opposite being true for the other parental chromosome (e.g. red and Long). This is referred to as repulsion or a trans arrangement. The phenotype here would still be purple and long but a test cross of this individual with the recessive parent would produce progeny with much greater proportion of the two crossover phenotypes. While such a problem may not seem likely from this example, unfavourable repulsion linkages do appear when breeding for disease resistance in some crops.

The two possible arrangements, cis and trans, of alleles in a double heterozygote are referred to as gametic phases, and phasing is the process of determining which of the two is present in a given individual.

When two genes are located on the same chromosome, the chance of a crossover producing recombination between the genes is related to the distance between the two genes. Thus, the use of recombination frequencies has been used to develop linkage maps or genetic maps.

However, it is important to note that recombination frequency tends to underestimate the distance between two linked genes. This is because as the two genes are located farther apart, the chance of double or even number of crossovers between them also increases. Double or even number of crossovers between the two genes results in them being cosegregated to the same gamete, yielding a parental progeny instead of the expected recombinant progeny. As mentioned above, the Kosambi and Haldane transformations attempt to correct for multiple crossovers.

Variation of Recombination Frequency

While recombination of chromosomes is an essential process during meiosis, there is a large range of frequency of cross overs across organisms and within species. Sexually dimorphic rates of recombination are termed heterochiasmy, and are observed more often than a common rate between male and females. In mammals, females often have a higher rate of recombination compared to males. It is theorised that there are unique selections acting or meiotic drivers which influence the difference in rates. The difference in rates may also reflect the vastly different environments and conditions of meiosis in oogenesis and spermatogenesis.

Meiosis Indicators

With very large pedigrees or with very dense genetic marker data, such as from whole-genome sequencing, it is possible to precisely locate recombinations. With this type of genetic analysis, a meiosis indicator is assigned to each position of the genome for each meiosis in a pedigree. The indicator indicates which copy of the parental chromosome contributes to the transmitted gamete at that position. For example, if the allele from the 'first' copy of the parental chromosome is transmitted, a '0' might be assigned to that meiosis. If the allele from the 'second' copy of the parental chromosome is transmitted, a '1' would be assigned to that meiosis. The two alleles in the parent came, one each, from two grandparents. These indicators are then used to determine identical-by-descent (IBD) states or inheritance states, which are in turn used to identify genes responsible for diseases.

Chromosomal Crossover

Chromosomal crossover (or crossing over) is the exchange of genetic material between two homologous chromosomes non-sister chromatids that results in recombinant chromosomes during sexual reproduction. It is one of the final phases of genetic recombination, which occurs in the pachytene stage of prophase I of meiosis during a process called synapsis. Synapsis begins before the synaptonemal complex develops and is not completed until near the end of prophase I. Crossover usually occurs when matching regions on matching chromosomes break and then reconnect to the other chromosome.

Crossing over was described, in theory, by Thomas Hunt Morgan. He relied on the discovery of Frans Alfons Janssens who described the phenomenon in 1909 and had called it "chiasmatypie". The term chiasma is linked, if not identical, to chromosomal crossover. Morgan immediately saw the great importance of Janssens' cytological interpretation of chiasmata to the experimental results of his research on the heredity of Drosophila. The physical basis of crossing over was first demonstrated by Harriet Creighton and Barbara McClintock in 1931.

The linked frequency of crossing over between two gene loci (markers) is the *crossing-over value*. For fixed set of genetic and environmental conditions, recombination in a particular region of a linkage structure (chromosome) tends to be constant and the same is then true for the crossing-over value which is used in the production of genetic maps.

There are two popular and overlapping theories that explain the origins of crossing-over, coming from the different theories on the origin of meiosis. The first theory rests upon the idea that meiosis evolved as another method of DNA repair, and thus crossing-over is a novel way to replace possibly damaged sections of DNA. The second theory comes from the idea that meiosis evolved from bacterial transformation, with the function of propagating diversity. In 1931, Barbara McClintock discovered a triploid maize plant. She made key findings regarding corn's karyotype, including the size and shape of the chromosomes. McClintock used the prophase and metaphase stages of mitosis to describe the morphology of corn's chromosomes, and later showed the first ever cytological demonstration of crossing over in meiosis. Working with student Harriet Creighton, McClintock also made significant contributions to the early understanding of codependency of linked genes.

DNA Repair Theory

Crossing over and DNA repair are very similar processes, which utilize many of the same protein complexes. In her report, "The Significance of Responses of the Genome to Challenge", McClintock studied corn to show how corn's genome would change itself to overcome threats to its survival. She used 450 self-pollinated plants that received from each parent a chromosome with a ruptured end. She used modified patterns of gene expression on different sectors of leaves of her corn plants show that transposable elements ("controlling elements") hide in the genome, and their mobility allows them to alter the action of genes at different loci. These elements can also restructure the genome, anywhere from a few nucleotides to whole segments of chromosome. Recombinases and primases lay a foundation of nucleotides along the DNA sequence. One such particular protein complex that is conserved between processes is RAD51, a well conserved recombinase protein that has been shown to be crucial in DNA repair as well as cross over. Several other genes in *D. melanogaster* have been linked as well to both processes, by showing that mutants at these specific loci cannot undergo DNA repair or crossing over. Such genes include mei-41, mei-9, hdm, spnA, and brca2. This large group of conserved genes between processes supports the theory of a close evolutionary relationship. Furthermore, DNA repair and crossover have been found to favor similar regions on chromosomes. In an experiment using radiation hybrid mapping on wheat's (*Triticum aestivum L.*) 3B chromosome, crossing over and DNA repair were found to occur predominantly in the same regions. Furthermore, crossing over has been correlated to occur in response to stressful, and likely DNA damaging, conditions.

Links to Bacterial Transformation

The process of bacterial transformation also shares many similarities with chromosomal cross over, particularly in the formation of overhangs on the sides of the broken DNA strand, allowing for the annealing of a new strand. Bacterial transformation itself has been linked to DNA repair many times. The second theory comes from the idea that meiosis evolved from bacterial transformation, with the function of propagating genetic diversity. Thus, this evidence suggests that it is a question of whether cross over is linked to DNA repair or bacterial transformation, as the two do not appear to be mutually exclusive. It is likely that crossing over may have evolved from bacterial transformation, which in turn developed from DNA repair, thus explaining the links between all three processes.

Chemistry

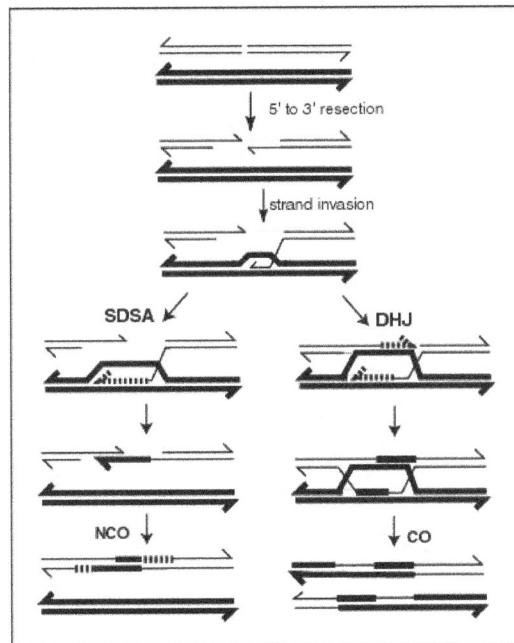

A current model of meiotic recombination, initiated by a double-strand break or gap, followed by pairing with a homologous chromosome and strand invasion to initiate the recombinational repair process. Repair of the gap can lead to crossover (CO) or non-crossover (NCO) of the flanking regions. CO recombination is thought to occur by the Double Holliday Junction (DHJ) model, illustrated on the right, above. NCO recombinants are thought to occur primarily by the Synthesis Dependent Strand Annealing (SDSA) model, illustrated on the left, above. Most recombination events appear to be the SDSA type.

Meiotic recombination may be initiated by double-stranded breaks that are introduced into the DNA by exposure to DNA damaging agents or the Spo11 protein. One or more exonucleases then digest the 5' ends generated by the double-stranded breaks to produce 3' single-stranded DNA tails. The meiosis-specific recombinase Dmc1 and the general recombinase Rad51 coat the single-stranded DNA to form nucleoprotein filaments. The recombinases catalyze invasion of the opposite chromatid by the single-stranded DNA from one end of the break. Next, the 3' end of the invading DNA primes DNA synthesis, causing displacement of the complementary strand, which subsequently anneals to the single-stranded DNA generated from the other end of the initial double-stranded break. The structure that results is a *cross-strand exchange*, also known as a Holliday junction. The contact between two chromatids that will soon undergo crossing-over is known as a *chiasma*. The Holliday junction is a tetrahedral structure which can be 'pulled' by other recombinases, moving it along the four-stranded structure.

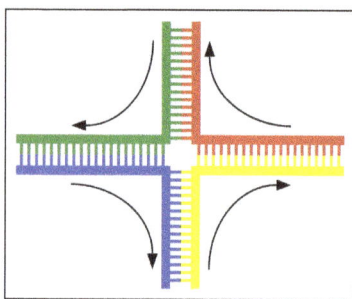

Holliday Junction.

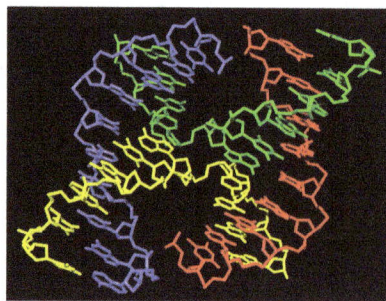

Molecular structure of a Holliday junction.

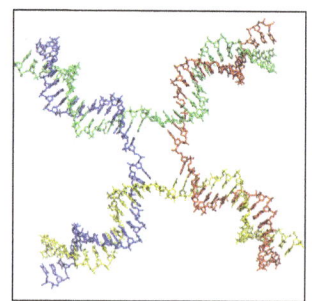

Molecular structure of a Holliday junction.

MSH4 and MSH5

The MSH4 and MSH5 proteins form a hetero-oligomeric structure (heterodimer) in yeast and humans. In the yeast *Saccharomyces cerevisiae* MSH4 and MSH5 act specifically to facilitate crossovers between homologous chromosomes during meiosis. The MSH4/MSH5 complex binds and stabilizes double Holliday junctions and promotes their resolution into crossover products. An MSH4 hypomorphic (partially functional) mutant of *S. cerevisiae* showed a 30% genome wide reduction in crossover numbers, and a large number of meioses with non exchange chromosomes. Nevertheless, this mutant gave rise to spore viability patterns suggesting that segregation of non-exchange chromosomes occurred efficiently. Thus in *S. cerevisiae* proper segregation apparently does not entirely depend on crossovers between homologous pairs.

Chiasma

The grasshopper Melanoplus femur-rubrum was exposed to an acute dose of X-rays during each individual stage of meiosis, and chiasma frequency was measured. Irradiation during the leptotene-zygotene stages of meiosis (that is, prior to the pachytene period in which crossover recombination occurs) was found to increase subsequent chiasma frequency. Similarly, in the grasshopper

Chorthippus brunneus, exposure to X-irradiation during the zygotene-early pachytene stages caused a significant increase in mean cell chiasma frequency. Chiasma frequency was scored at the later diplotene-diakinesis stages of meiosis. These results suggest that X-rays induce DNA damages that are repaired by a crossover pathway leading to chiasma formation.

Consequences:

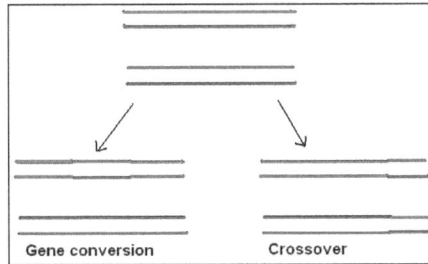

The difference between gene conversion and chromosomal crossover.

In most eukaryotes, a cell carries two versions of each gene, each referred to as an allele. Each parent passes on one allele to each offspring. An individual gamete inherits a complete haploid complement of alleles on chromosomes that are independently selected from each pair of chromatids lined up on the metaphase plate. Without recombination, all alleles for those genes linked together on the same chromosome would be inherited together. Meiotic recombination allows a more independent segregation between the two alleles that occupy the positions of single genes, as recombination shuffles the allele content between homologous chromosomes.

Recombination results in a new arrangement of maternal and paternal alleles on the same chromosome. Although the same genes appear in the same order, some alleles are different. In this way, it is theoretically possible to have any combination of parental alleles in an offspring, and the fact that two alleles appear together in one offspring does not have any influence on the statistical probability that another offspring will have the same combination. This principle of "independent assortment" of genes is fundamental to genetic inheritance. However, the frequency of recombination is actually not the same for all gene combinations. This leads to the notion of "genetic distance", which is a measure of recombination frequency averaged over a (suitably large) sample of pedigrees. Loosely speaking, one may say that this is because recombination is greatly influenced by the proximity of one gene to another. If two genes are located close together on a chromosome, the likelihood that a recombination event will separate these two genes is less than if they were farther apart. Genetic linkage describes the tendency of genes to be inherited together as a result of their location on the same chromosome. Linkage disequilibrium describes a situation in which some combinations of genes or genetic markers occur more or less frequently in a population than would be expected from their distances apart. This concept is applied when searching for a gene that may cause a particular disease. This is done by comparing the occurrence of a specific DNA sequence with the appearance of a disease. When a high correlation between the two is found, it is likely that the appropriate gene sequence is really closer.

Non-homologous Crossover

Crossovers typically occur between homologous regions of matching chromosomes, but similarities in sequence and other factors can result in mismatched alignments. Most DNA is composed of base pair sequences repeated very large numbers of times. These repetitious segments, often referred to as satellites, are fairly homogenous among a species. During DNA replication, each strand of DNA is used as a template for the creation of new strands using a partially-conserved mechanism; proper functioning of this process results in two identical, paired chromosomes, often called sisters. Sister chromatid crossover events are known to occur at a rate of several crossover events per cell per division in eukaryotes. Most of these events involve an exchange of equal amounts of genetic information, but unequal exchanges may occur due to sequence mismatch. These are referred to by a variety of names, including non-homologous crossover, unequal crossover, and unbalanced recombination, and result in an insertion or deletion of genetic information into the chromosome. While rare compared to homologous crossover events, these mutations are drastic, affecting many loci at the same time. They are considered the main driver behind the generation of gene duplications and are a general source of mutation within the genome.

The specific causes of non-homologous crossover events are unknown, but several influential factors are known to increase the likelihood of an unequal crossover. One common vector leading to unbalanced recombination is the repair of double-strand breaks (DSBs). DSBs are often repaired using homology directed repair, a process which involves invasion of a template strand by the DSB strand. Nearby homologous regions of the template strand are often used for repair, which can give rise to either insertions or deletions in the genome if a non-homologous but complementary part of the template strand is used. Sequence similarity is a major player in crossover – crossover events are more likely to occur in long regions of close identity on a gene. This means that any section of the genome with long sections of repetitive DNA is prone to crossover events.

The presence of transposable elements is another influential element of non-homologous crossover. Repetitive regions of code characterize transposable elements; complementary but non-homologous regions are ubiquitous within transposons. Because chromosomal regions composed of transposons have large quantities of identical, repetitive code in a condensed space, it is thought that transposon regions undergoing a crossover event are more prone to erroneous complementary match-up; that is to say, a section of a chromosome containing a lot of identical sequences, should it undergo a crossover event, is less certain to match up with a perfectly homologous section of complementary code and more prone to binding with a section of code on a slightly different part of the chromosome. This results in unbalanced recombination, as genetic information may be either inserted or deleted into the new chromosome, depending on where the recombination occurred.

While the motivating factors behind unequal recombination remain obscure, elements of the physical mechanism have been elucidated. Mismatch repair (MMR) proteins, for instance, are a well-known regulatory family of proteins, responsible for regulating mismatched sequences of DNA during replication and escape regulation. The operative goal of MMRs is the restoration of the parental genotype. One class of MMR in particular, MutSβ, is known to initiate the correction of insertion-deletion mismatches of up to 16 nucleotides. Little is known about the excision process in eukaryotes, but E. coli excisions involve the cleaving of a nick on either the 5' or 3' strand, after which DNA helicase and DNA polymerase III bind and generate single-stranded proteins, which are digested by exonucleases and attached to the strand by ligase. Multiple MMR pathways have been implicated in the maintenance of complex organism genome stability, and any of many possible malfunctions in the MMR pathway result in DNA editing and correction errors. Therefore, while it is not certain precisely what mechanisms lead to errors of non-homologous crossover, it is extremely likely that the MMR pathway is involved.

Genetic Analysis

Genetic analysis is the overall process of studying and researching in fields of science that involve genetics and molecular biology. There are a number of applications that are developed from this research, and these are also considered parts of the process. The base system of analysis revolves around general genetics. Basic studies include identification of genes and inherited disorders. This research has been conducted for centuries on both a large-scale physical observation basis and on a more microscopic scale. Genetic analysis can be used generally to describe methods both used in and resulting from the sciences of genetics and molecular biology, or to applications resulting from this research.

Genetic analysis may be done to identify genetic/inherited disorders and also to make a differential diagnosis in certain somatic diseases such as cancer. Genetic analyses of cancer include detection of mutations, fusion genes, and DNA copy number changes.

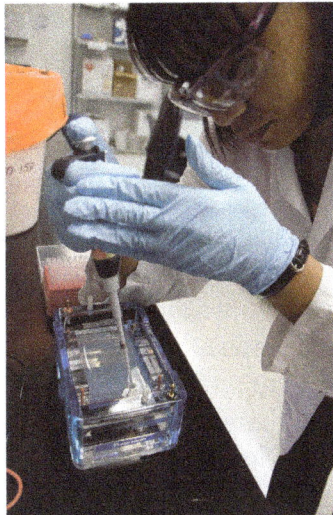

FDA microbiologist prepares DNA samples
for gel electrophoresis analysis.

Much of the research that set the foundation of genetic analysis began in prehistoric times. Early humans found that they could practice selective breeding to improve crops and animals. They also identified inherited traits in humans that were eliminated over the years. The many genetic analyses gradually evolved over time.

Mendelian Research

Modern genetic analysis began in the mid-1800s with research conducted by Gregor Mendel. Mendel, who is known as the "father of modern genetics", was inspired to study variation in plants. Between 1856 and 1863, Mendel cultivated and tested some 29,000 pea plants (i.e., Pisum sativum). This study showed that one in four pea plants had purebred recessive alleles, two out of four were hybrid and one out of four were purebred dominant. His experiments led him to make two generalizations, the Law of Segregation and the Law of Independent Assortment, which later became known as Mendel's Laws of Inheritance. Lacking the basic understanding of heredity,

Mendel observed various organisms and first utilized genetic analysis to find that traits were inherited from parents and those traits could vary between children. Later, it was found that units within each cell are responsible for these traits. These units are called genes. Each gene is defined by a series of amino acids that create proteins responsible for genetic traits.

Various Types of Genetic Analysis

Genetic analyses include molecular technologies such as PCR, RT-PCR, DNA sequencing, and DNA microarrays, and cytogenetic methods such as karyotyping and fluorescence in situ hybridisation.

Electrophoresis apparatus.

DNA Sequencing

DNA sequencing is essential to the applications of genetic analysis. This process is used to determine the order of nucleotide bases. Each molecule of DNA is made from adenine, guanine, cytosine and thymine, which determine what function the genes will possess. This was first discovered during the 1970s. DNA sequencing encompasses biochemical methods for determining the order of the nucleotide bases, adenine, guanine, cytosine, and thymine, in a DNA oligonucleotide. By generating a DNA sequence for a particular organism, you are determining the patterns that make up genetic traits and in some cases behaviors.

Sequencing methods have evolved from relatively laborious gel-based procedures to modern automated protocols based on dye labelling and detection in capillary electrophoresis that permit rapid large-scale sequencing of genomes and transcriptomes. Knowledge of DNA sequences of genes and other parts of the genome of organisms has become indispensable for basic research studying biological processes, as well as in applied fields such as diagnostic or forensic research. The advent of DNA sequencing has significantly accelerated biological research and discovery.

Cytogenetics

Cytogenetics is a branch of genetics that is concerned with the study of the structure and function of the cell, especially the chromosomes. Polymerase chain reaction studies the amplification of DNA. Because of the close analysis of chromosomes in cytogenetics, abnormalities are more readily seen and diagnosed.

Karyotyping

A karyotype is the number and appearance of chromosomes in the nucleus of a eukaryotic cell. The term is also used for the complete set of chromosomes in a species, or an individual organism.

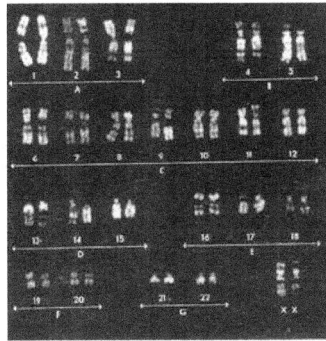

Karyotype of chromosomes.

Karyotypes describe the number of chromosomes, and what they look like under a light microscope. Attention is paid to their length, the position of the centromeres, banding pattern, any differences between the sex chromosomes, and any other physical characteristics. Karyotyping uses a system of studying chromosomes to identify genetic abnormalities and evolutionary changes in the past.

DNA Microarrays

A DNA microarray is a collection of microscopic DNA spots attached to a solid surface. Scientists use DNA microarrays to measure the expression levels of large numbers of genes simultaneously or to genotype multiple regions of a genome. When a gene is expressed in a cell, it generates messenger RNA (mRNA). Overexpressed genes generate more mRNA than underexpressed genes. This can be detected on the microarray Since an array can contain tens of thousands of probes, a microarray experiment can accomplish many genetic tests in parallel. Therefore, arrays have dramatically accelerated many types of investigations.

PCR

The polymerase chain reaction (PCR) is a biochemical technology in molecular biology to amplify a single or a few copies of a piece of DNA across several orders of magnitude, generating thousands to millions of copies of a particular DNA sequence. PCR is now a common and often indispensable technique used in medical and biological research labs for a variety of applications. These include DNA cloning for sequencing, DNA-based phylogeny, or functional analysis of genes; the diagnosis of hereditary diseases; the identification of genetic fingerprints (used in forensic sciences and paternity testing); and the detection and diagnosis of infectious diseases.

Practical Application

Cancer Breakthrough

Numerous practical advancements have been made in the field of genetics and molecular biology through the processes of genetic analysis. One of the most prevalent advancements during the late 20th and early 21st centuries is a greater understanding of cancer's link to genetics. By identifying which genes in the cancer cells are working abnormally, doctors can better diagnose and treat cancers.

Possibilities

This research has been able to identify the concepts of genetic mutations, fusion genes and changes in DNA copy numbers, and advances are made in the field every day. Much of these applications have led to new types of sciences that use the foundations of genetic analysis. Reverse genetics uses the methods to determine what is missing in a genetic code or what can be added to change that code. Genetic linkage studies analyze the spatial arrangements of genes and chromosomes. There have also been studies to determine the legal and social and moral effects of the increase of genetic analysis. Genetic analysis may be done to identify genetic/inherited disorders and also to make a differential diagnosis in certain somatic diseases such as cancer. Genetic analyses of cancer include detection of mutations, fusion genes, and DNA copy number changes.

Gene Mapping

Gene mapping describes the methods used to identify the locus of a gene and the distances between genes.

The essence of all genome mapping is to place a collection of molecular markers onto their respective positions on the genome. Molecular markers come in all forms. Genes can be viewed as one special type of genetic markers in the construction of genome maps, and mapped the same way as any other markers.

Genetic Mapping vs. Physical Mapping

There are two distinctive types of "Maps" used in the field of genome mapping: genetic maps and physical maps. While both maps are a collection of genetic markers and gene loci, genetic maps' distances are based on the genetic linkage information, while physical maps use actual physical distances usually measured in number of base pairs. While the physical map could be a more "accurate" representation of the genome, genetic maps often offer insights into the nature of different regions of the chromosome, e.g. the genetic distance to physical distance ratio varies greatly at different genomic regions which reflects different recombination rates, and such rate is often indicative of euchromatic (usually gene-rich) vs heterochromatic (usually gene poor) regions of the genome.

Gene Mapping

Researchers begin a genetic map by collecting samples of blood or tissue from family members that carry a prominent disease or trait and family members that don't. Scientists then isolate DNA from the samples and closely examine it, looking for unique patterns in the DNA of the family members who do carry the disease that the DNA of those who don't carry the disease don't have. These unique molecular patterns in the DNA are referred to as polymorphisms, or markers.

The first steps of building a genetic map are the development of genetic markers and a mapping population. The closer two markers are on the chromosome, the more likely they are to be passed on to the next generation together. Therefore, the "co-segregation" patterns of all markers can be used to reconstruct their order. With this in mind, the genotypes of each genetic marker are recorded for both parents and each individual in the following generations. The quality of the genetic maps is largely dependent upon these factors: the number of genetic markers on the map and the size of the mapping population. The two factors are interlinked, as a larger mapping population could increase the "resolution" of the map and prevent the map being "saturated".

In gene mapping, any sequence feature that can be faithfully distinguished from the two parents can be used as a genetic marker. Genes, in this regard, are represented by "traits" that can be faithfully distinguished between two parents. Their linkage with other genetic markers are calculated in the same way as if they are common markers and the actual gene loci are then bracketed in a region between the two nearest neighbouring markers. The entire process is then repeated by looking at more markers which target that region to map the gene neighbourhood to a higher resolution until a specific causative locus can be identified. This process is often referred to as "positional cloning", and it is used extensively in the study of plant species The great advantage of genetic mapping is that it can identify the relative position of genes based solely on their phenotypic effect.

Physical Mapping

Since actual base-pair distances are generally hard or impossible to directly measure, physical maps are actually constructed by first shattering the genome into hierarchically smaller pieces. By characterizing each single piece and assembling back together, the overlapping path or "tiling path" of these small fragments would allow researchers to infer physical distances between genomic features. The fragmentation of the genome can be achieved by restriction enzyme cutting or by physically shattering the genome by processes like sonication. Once cut, the DNA fragments are separated by electrophoresis. The resulting pattern of DNA migration (i.e. its genetic fingerprint) is used to identify what stretch of DNA is in the clone. By analyzing the fingerprints, contigs are assembled by automated (FPC) or manual means (pathfinders) into overlapping DNA stretches. Now a good choice of clones can be made to efficiently sequence the clones to determine the DNA sequence of the organism under study.

In physical mapping, there are no direct ways of marking up a specific gene since the mapping does not include any information that concerns traits and functions. Genetic markers can be linked to a physical map by processes like in situ hybridization. By this approach, physical map contigs can be "anchored" onto a genetic map. The clones used in the physical map contigs can then be sequenced on a local scale to help new genetic marker design and identification of the causative loci.

Macrorestriction is a type of physical mapping wherein the high molecular weight DNA is digested with a restriction enzyme having a low number of restriction sites.

There are alternative ways to determine how DNA in a group of clones overlaps without completely sequencing the clones. Once the map is determined, the clones can be used as a resource to efficiently contain large stretches of the genome. This type of mapping is more accurate than genetic maps.

Genome Sequencing

Genome sequencing is sometimes mistakenly referred to as "genome mapping" by non-biologists. The process of "shotgun sequencing" resembles the process of physical mapping: it shatters the genome into small fragments, characterizes each fragment, then puts them back together (more recent sequencing technologies are drastically different). While the scope, purpose and process are totally different, a genome assembly can be viewed as the "ultimate" form of physical map, in that it provides in a much better way all the information that a traditional physical map can offer.

Use

Identification of genes is usually the first step in understanding a genome of a species; mapping of the gene is usually the first step of identification of the gene. Gene mapping is usually the starting point of many important downstream studies.

Disease Association

The process to identify a genetic element that is responsible for a disease is also referred to as "mapping". If the locus in which the search is performed is already considerably constrained, the search is called the fine mapping of a gene. This information is derived from the investigation of disease manifestations in large families (genetic linkage) or from populations-based genetic association studies.

Medical Genetics

Medical genetics is the branch of medicine that involves the diagnosis and management of hereditary disorders. Medical genetics differs from human genetics in that human genetics is a field of scientific research that may or may not apply to medicine, while medical genetics refers to the application of genetics to medical care. For example, research on the causes and inheritance of genetic disorders would be considered within both human genetics and medical genetics, while the diagnosis, management, and counselling people with genetic disorders would be considered part of medical genetics.

In contrast, the study of typically non-medical phenotypes such as the genetics of eye color would be considered part of human genetics, but not necessarily relevant to medical genetics (except in situations such as albinism). *Genetic medicine* is a newer term for medical genetics and incorporates areas such as gene therapy, personalized medicine, and the rapidly emerging new medical specialty, predictive medicine.

Scope

Medical genetics encompasses many different areas, including clinical practice of physicians, genetic counselors, and nutritionists, clinical diagnostic laboratory activities, and research into the causes and inheritance of genetic disorders. Examples of conditions that fall within the scope of medical genetics include birth defects and dysmorphology, mental retardation, autism, mitochondrial disorders, skeletal dysplasia, connective tissue disorders, cancer genetics, teratogens, and prenatal diagnosis. Medical genetics is increasingly becoming relevant to many common diseases. Overlaps with other medical specialties are beginning to emerge, as recent advances in genetics are revealing etiologies for neurologic, endocrine, cardiovascular, pulmonary, ophthalmologic, renal, psychiatric, and dermatologic conditions. The medical genetics community is increasingly involved with individuals who have undertaken elective genetic and genomic testing.

Subspecialties

In some ways, many of the individual fields within medical genetics are hybrids between clinical care and research. This is due in part to recent advances in science and technology that have enabled an unprecedented understanding of genetic disorders.

Clinical Genetics

Clinical genetics is the practice of clinical medicine with particular attention to hereditary disorders. Referrals are made to genetics clinics for a variety of reasons, including birth defects, developmental delay, autism, epilepsy, short stature, and many others. Examples of genetic syndromes that are commonly seen in the genetics clinic include chromosomal rearrangements, Down syndrome, DiGeorge syndrome, Fragile X syndrome, Marfan syndrome, Neurofibromatosis, Turner syndrome, and Williams syndrome.

In the United States, Doctors who practice clinical genetics are accredited by the American Board of Medical Genetics and Genomics (ABMGG). In order to become a board-certified practitioner of Clinical Genetics, a physician must complete a minimum of 24 months of training in a program accredited by the ABMGG. Individuals seeking acceptance into clinical genetics training programs must hold an M.D. or D.O. degree (or their equivalent) and have completed a minimum of 24 months of training in an ACGME-accredited residency program in internal medicine, pediatrics, obstetrics and gynecology, or other medical specialty.

Metabolic/Biochemical Genetics

Metabolic (or biochemical) genetics involves the diagnosis and management of inborn errors of metabolism in which patients have enzymatic deficiencies that perturb biochemical pathways involved in metabolism of carbohydrates, amino acids, and lipids. Examples of metabolic disorders include galactosemia, glycogen storage disease, lysosomal storage disorders, metabolic acidosis, peroxisomal disorders, phenylketonuria, and urea cycle disorders.

Cytogenetics

Cytogenetics is the study of chromosomes and chromosome abnormalities. While cytogenetics historically relied on microscopy to analyze chromosomes, new molecular technologies such as array comparative genomic hybridization are now becoming widely used. Examples of chromosome abnormalities include aneuploidy, chromosomal rearrangements, and genomic deletion/duplication disorders.

Molecular Genetics

Molecular genetics involves the discovery of and laboratory testing for DNA mutations that underlie many single gene disorders. Examples of single gene disorders include achondroplasia, cystic fibrosis, Duchenne muscular dystrophy, hereditary breast cancer (BRCA1/2), Huntington disease, Marfan syndrome, Noonan syndrome, and Rett syndrome. Molecular tests are also used in the diagnosis of syndromes involving epigenetic abnormalities, such as Angelman syndrome, Beckwith-Wiedemann syndrome, Prader-willi syndrome, and uniparental disomy.

Mitochondrial Genetics

Mitochondrial genetics concerns the diagnosis and management of mitochondrial disorders, which have a molecular basis but often result in biochemical abnormalities due to deficient energy production.

There exists some overlap between medical genetic diagnostic laboratories and molecular pathology.

Genetic Counseling

Genetic counseling is the process of providing information about genetic conditions, diagnostic testing, and risks in other family members, within the framework of nondirective counseling. Genetic counselors are non-physician members of the medical genetics team who specialize in family risk assessment and counseling of patients regarding genetic disorders. The precise role of the genetic counselor varies somewhat depending on the disorder.

Current Practice

The clinical setting in which patients are evaluated determines the scope of practice, diagnostic, and therapeutic interventions. For the purposes of general discussion, the typical encounters between patients and genetic practitioners may involve:

- Referral to an out-patient genetics clinic (pediatric, adult, or combined) or an in-hospital consultation, most often for diagnostic evaluation.

- Specialty genetics clinics focusing on management of inborn errors of metabolism, skeletal dysplasia, or lysosomal storage diseases.

- Referral for counseling in a prenatal genetics clinic to discuss risks to the pregnancy (advanced maternal age, teratogen exposure, family history of a genetic disease), test results (abnormal maternal serum screen, abnormal ultrasound), and/or options for prenatal diagnosis (typically non-invasive prenatal screening, diagnostic amniocentesis or chorionic villus sampling).

- Multidisciplinary specialty clinics that include a clinical geneticist or genetic counselor (cancer genetics, cardiovascular genetics, craniofacial or cleft lip/palate, hearing loss clinics, muscular dystrophy/neurodegenerative disorder clinics).

Diagnostic Evaluation

Each patient will undergo a diagnostic evaluation tailored to their own particular presenting signs and symptoms. The geneticist will establish a differential diagnosis and recommend appropriate testing. These tests might evaluate for chromosomal disorders, inborn errors of metabolism, or single gene disorders.

Chromosome Studies

Chromosome studies are used in the general genetics clinic to determine a cause for developmental delay/mental retardation, birth defects, dysmorphic features, and/or autism. Chromosome analysis is also performed in the prenatal setting to determine whether a fetus is affected with aneuploidy or other chromosome rearrangements. Finally, chromosome abnormalities are often detected in cancer samples. A large number of different methods have been developed for chromosome analysis:

- Chromosome analysis using a karyotype involves special stains that generate light and dark bands, allowing identification of each chromosome under a microscope.

- Fluorescence in situ hybridization (FISH) involves fluorescent labeling of probes that bind to specific DNA sequences, used for identifying aneuploidy, genomic deletions or duplications, characterizing chromosomal translocations and determining the origin of ring chromosomes.

- Chromosome painting is a technique that uses fluorescent probes specific for each chromosome to differentially label each chromosome. This technique is more often used in cancer cytogenetics, where complex chromosome rearrangements can occur.

- Array comparative genomic hybridization is a newer molecular technique that involves hybridization of an individual DNA sample to a glass slide or microarray chip containing molecular probes (ranging from large ~200kb bacterial artificial chromosomes to small oligonucleotides) that represent unique regions of the genome. This method is particularly sensitive for detection of genomic gains or losses across the genome but does not detect balanced translocations or distinguish the location of duplicated genetic material (for example, a tandem duplication versus an insertional duplication).

Basic Metabolic Studies

Biochemical studies are performed to screen for imbalances of metabolites in the bodily fluid, usually the blood (plasma/serum) or urine, but also in cerebrospinal fluid (CSF). Specific tests of enzyme function (either in leukocytes, skin fibroblasts, liver, or muscle) are also employed under certain circumstances. In the US, the newborn screen incorporates biochemical tests to screen for treatable conditions such as galactosemia and phenylketonuria (PKU). Patients suspected to have a metabolic condition might undergo the following tests:

- Quantitative amino acid analysis is typically performed using the ninhydrin reaction, followed by liquid chromatography to measure the amount of amino acid in the sample (either urine, plasma/serum, or CSF). Measurement of amino acids in plasma or serum is used in the evaluation of disorders of amino acid metabolism such as urea cycle disorders, maple syrup urine disease, and PKU. Measurement of amino acids in urine can be useful in the diagnosis of cystinuria or renal Fanconi syndrome as can be seen in cystinosis.

- Urine organic acid analysis can be either performed using quantitative or qualitative methods, but in either case the test is used to detect the excretion of abnormal organic acids. These compounds are normally produced during bodily metabolism of amino acids and odd-chain fatty acids, but accumulate in patients with certain metabolic conditions.

- The acylcarnitine combination profile detects compounds such as organic acids and fatty acids conjugated to carnitine. The test is used for detection of disorders involving fatty acid metabolism, including MCAD.

- Pyruvate and lactate are byproducts of normal metabolism, particularly during anaerobic metabolism. These compounds normally accumulate during exercise or ischemia, but are also elevated in patients with disorders of pyruvate metabolism or mitochondrial disorders.

- Ammonia is an end product of amino acid metabolism and is converted in the liver to urea through a series of enzymatic reactions termed the urea cycle. Elevated ammonia can therefore be detected in patients with urea cycle disorders, as well as other conditions involving liver failure.

- Enzyme testing is performed for a wide range of metabolic disorders to confirm a diagnosis suspected based on screening tests.

Molecular Studies

- DNA sequencing is used to directly analyze the genomic DNA sequence of a particular gene. In general, only the parts of the gene that code for the expressed protein (exons) and small amounts of the flanking untranslated regions and introns are analyzed. Therefore, although these tests are highly specific and sensitive, they do not routinely identify all of the mutations that could cause disease.

- DNA methylation analysis is used to diagnose certain genetic disorders that are caused by disruptions of epigenetic mechanisms such as genomic imprinting and uniparental disomy.

- Southern blotting is an early technique basic on detection of fragments of DNA separated by size through gel electrophoresis and detected using radiolabeled probes. This test was routinely used to detect deletions or duplications in conditions such as Duchenne muscular dystrophy but is being replaced by high-resolution array comparative genomic hybridization techniques. Southern blotting is still useful in the diagnosis of disorders caused by trinucleotide repeats.

Treatments

Each cell of the body contains the hereditary information (DNA) wrapped up in structures called chromosomes. Since genetic syndromes are typically the result of alterations of the chromosomes or genes, there is no treatment currently available that can correct the genetic alterations in every cell of the body. Therefore, there is currently no "cure" for genetic disorders. However, for many genetic syndromes there is treatment available to manage the symptoms. In some cases, particularly inborn errors of metabolism, the mechanism of disease is well understood and offers the potential for dietary and medical management to prevent or reduce the long-term complications. In other cases, infusion therapy is used to replace the missing enzyme. Current research is actively seeking to use gene therapy or other new medications to treat specific genetic disorders.

Management of Metabolic Disorders

In general, metabolic disorders arise from enzyme deficiencies that disrupt normal metabolic pathways. For instance, in the hypothetical example:

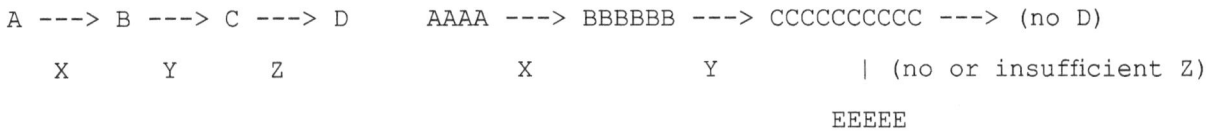

```
A ---> B ---> C ---> D      AAAA ---> BBBBBB ---> CCCCCCCCCC ---> (no D)

  X      Y     Z              X          Y        | (no or insufficient Z)

                                                EEEEE
```

Compound "A" is metabolized to "B" by enzyme "X", compound "B" is metabolized to "C" by enzyme "Y", and compound "C" is metabolized to "D" by enzyme "Z". If enzyme "Z" is missing, compound "D" will be missing, while compounds "A", "B", and "C" will build up. The pathogenesis of this particular condition could result from lack of compound "D", if it is critical for some cellular function, or from toxicity due to excess "A", "B", and/or "C", or from toxicity due to the excess of "E" which is normally only present in small amounts and only accumulates when "C" is in excess. Treatment of the metabolic disorder could be achieved through dietary supplementation of compound "D" and dietary restriction of compounds "A", "B", and/or "C" or by treatment with a medication that promoted disposal of excess "A", "B", "C" or "E". Another approach that can be taken is enzyme replacement therapy, in which a patient is given an infusion of the missing enzyme "Z" or cofactor therapy to increase the efficacy of any residual "Z" activity.

- Diet: Dietary restriction and supplementation are key measures taken in several well-known metabolic disorders, including galactosemia, phenylketonuria (PKU), maple syrup urine disease, organic acidurias and urea cycle disorders. Such restrictive diets can be difficult for the patient and family to maintain, and require close consultation with a nutritionist who has special experience in metabolic disorders. The composition of the diet will change depending on the caloric needs of the growing child and special attention is needed during a pregnancy if a woman is affected with one of these disorders.

- Medication: Medical approaches include enhancement of residual enzyme activity (in cases where the enzyme is made but is not functioning properly), inhibition of other enzymes in the biochemical pathway to prevent buildup of a toxic compound, or diversion of a toxic compound to another form that can be excreted. Examples include the use of high doses of pyridoxine (vitamin B6) in some patients with homocystinuria to boost the activity of the residual cystathione synthase enzyme, administration of biotin to restore activity of several enzymes affected by deficiency of biotinidase, treatment with NTBC in Tyrosinemia to inhibit the production of succinylacetone which causes liver toxicity, and the use of sodium benzoate to decrease ammonia build-up in urea cycle disorders.

- Enzyme replacement therapy: Certain lysosomal storage diseases are treated with infusions of a recombinant enzyme (produced in a laboratory), which can reduce the accumulation of the compounds in various tissues. Examples include Gaucher disease, Fabry disease, Mucopolysaccharidoses and Glycogen storage disease type II. Such treatments are limited by the ability of the enzyme to reach the affected areas (the blood brain barrier prevents enzyme from reaching the brain, for example), and can sometimes be associated with allergic reactions. The long-term clinical effectiveness of enzyme replacement therapies vary widely among different disorders.

Other Examples:

- Angiotensin receptor blockers in Marfan syndrome & Loeys-Dietz.

- Bone marrow transplantation.

- Gene therapy.

Ethical, Legal and Social Implications

Genetic information provides a unique type of knowledge about an individual and his/her family, fundamentally different from a typically laboratory test that provides a "snapshot" of an individual's health status. The unique status of genetic information and inherited disease has a number of ramifications with regard to ethical, legal, and societal concerns.

On 19 March 2015, scientists urged a worldwide ban on clinical use of methods, particularly the use of CRISPR and zinc finger, to edit the human genome in a way that can be inherited. In April 2015 and April 2016, Chinese researchers reported results of basic research to edit the DNA of non-viable human embryos using CRISPR. In February 2016, British scientists were given permission by regulators to genetically modify human embryos by using CRISPR and related techniques on condition that the embryos were destroyed within seven days. In June 2016 the Dutch government was reported to be planning to follow suit with similar regulations which would specify a 14-day limit.

Epigenetics

Epigenetics is the study of heritable phenotype changes that do not involve alterations in the DNA sequence. Epigenetics most often denotes changes that affect gene activity and expression, but can also be used to describe any heritable phenotypic change. Such effects on cellular and physiological phenotypic traits may result from external or environmental factors, or be part of normal development. The standard definition of epigenetics requires these alterations to be heritable, in the progeny of either cells or organisms.

The term also refers to the changes themselves: functionally relevant changes to the genome that do not involve a change in the nucleotide sequence. Examples of mechanisms that produce such changes are DNA methylation and histone modification, each of which alters how genes are expressed without altering the underlying DNA sequence. Gene expression can be controlled through the action of repressor proteins that attach to silencer regions of the DNA. These epigenetic changes may last through cell divisions for the duration of the cell's life, and may also last for multiple generations even though they do not involve changes in the underlying DNA sequence of the organism; instead, non-genetic factors cause the organism's genes to behave (or "express themselves") differently.

One example of an epigenetic change in eukaryotic biology is the process of cellular differentiation. During morphogenesis, totipotent stem cells become the various pluripotent cell lines of the embryo, which in turn become fully differentiated cells. In other words, as a single fertilized egg cell – the zygote – continues to divide, the resulting daughter cells change into all the different cell types in an organism, including neurons, muscle cells, epithelium, endothelium of blood vessels, etc., by activating some genes while inhibiting the expression of others.

Historically, some phenomena not necessarily heritable have also been described as epigenetic. For example, the term epigenetic has been used to describe any modification of chromosomal regions, especially histone modifications, whether or not these changes are heritable or associated with a phenotype. The consensus definition now requires a trait to be heritable for it to be considered epigenetic.

The term *epigenetics* in its contemporary usage emerged in the 1990s, but for some years has been used with somewhat variable meanings. A consensus definition of the concept of *epigenetic trait* as a "stably heritable phenotype resulting from changes in a chromosome without alterations in the DNA sequence" was formulated at a Cold Spring Harbor meeting in 2008, although alternate definitions that include non-heritable traits are still being used.

The term *epigenesis* has a generic meaning of "extra growth", and has been used in English since the 17th century.

Contemporary

Robin Holliday defined epigenetics as "the study of the mechanisms of temporal and spatial control of gene activity during the development of complex organisms." Thus, in its broadest sense, *epigenetic* can be used to describe anything other than DNA sequence that influences the development of an organism.

More recent usage of the word in biology follows stricter definitions. It is, as defined by Arthur Riggs and colleagues, "the study of mitotically and/or meiotically heritable changes in gene function that cannot be explained by changes in DNA sequence."

The term has also been used, however, to describe processes which have not been demonstrated to be heritable, such as some forms of histone modification; there are therefore attempts to redefine "epigenetics" in broader terms that would avoid the constraints of requiring heritability. For example, Adrian Bird defined epigenetics as "the structural adaptation of chromosomal regions so as to register, signal or perpetuate altered activity states." This definition would be inclusive of transient modifications associated with DNA repair or cell-cycle phases as well as stable changes maintained across multiple cell generations, but exclude others such as templating of membrane architecture and prions unless they impinge on chromosome function. Such redefinitions however are not universally accepted and are still subject to debate. The NIH "Roadmap Epigenomics Project", ongoing as of 2016, uses the following definition: "For purposes of this program, epigenetics refers to both heritable changes in gene activity and expression (in the progeny of cells or of individuals) and also stable, long-term alterations in the transcriptional potential of a cell that are not necessarily heritable." In 2008, a consensus definition of the epigenetic trait, a "stably heritable phenotype resulting from changes in a chromosome without alterations in the DNA sequence", was made at a Cold Spring Harbor meeting.

The similarity of the word to "genetics" has generated many parallel usages. The "epigenome" is a parallel to the word "genome", referring to the overall epigenetic state of a cell, and epigenomics refers to global analyses of epigenetic changes across the entire genome. The phrase "genetic code" has also been adapted – the "epigenetic code" has been used to describe the set of epigenetic features that create different phenotypes in different cells from the same underlying DNA sequence. Taken to its extreme, the "epigenetic code" could represent the total state of the cell, with the position of each molecule accounted for in an *epigenomic map*, a diagrammatic representation of the gene expression, DNA methylation and histone modification status of a particular genomic region. More typically, the term is used in reference to systematic efforts to measure specific, relevant forms of epigenetic information such as the histone code or DNA methylation patterns.

Developmental Psychology

In a sense somewhat unrelated to its use in biological disciplines, the term "epigenetic" has also been used in developmental psychology to describe psychological development as the result of an ongoing, bi-directional interchange between heredity and the environment. Interactivist ideas of development have been discussed in various forms and under various names throughout the 19th and 20th centuries. An early version was proposed, among the founding statements in embryology, by Karl Ernst von Baer and popularized by Ernst Haeckel. A radical epigenetic view (physiological epigenesis) was developed by Paul Wintrebert. Another variation, probabilistic epigenesis, was presented by Gilbert Gottlieb in 2003. This view encompasses all of the possible developing factors on an organism and how they not only influence the organism and each other, but how the organism also influences its own development.

The developmental psychologist Erik Erikson wrote of an *epigenetic principle* in his 1968 book *Identity: Youth and Crisis*, encompassing the notion that we develop through an unfolding of our personality in predetermined stages, and that our environment and surrounding culture influence how we progress through these stages. This biological unfolding in relation to our socio-cultural settings is done in stages of psychosocial development, where "progress through each stage is in part determined by our success, or lack of success, in all the previous stages."

Molecular Basis

Epigenetic changes modify the activation of certain genes, but not the genetic code sequence of DNA. The microstructure (not code) of DNA itself or the associated chromatin proteins may be modified, causing activation or silencing. This mechanism enables differentiated cells in a multicellular organism to express only the genes that are necessary for their own activity. Epigenetic changes are preserved when cells divide. Most epigenetic changes only occur within the course of one individual organism's lifetime; however, these epigenetic changes can be transmitted to the organism's offspring through a process called transgenerational epigenetic inheritance. Moreover, if gene inactivation occurs in a sperm or egg cell that results in fertilization, this epigenetic modification may also be transferred to the next generation.

Specific epigenetic processes include paramutation, bookmarking, imprinting, gene silencing, X chromosome inactivation, position effect, DNA methylation reprogramming, transvection, maternal effects, the progress of carcinogenesis, many effects of teratogens, regulation of histone modifications and heterochromatin, and technical limitations affecting parthenogenesis and cloning.

DNA Damage

DNA damage can also cause epigenetic changes. DNA damage is very frequent, occurring on average about 60,000 times a day per cell of the human body. These damages are largely repaired, but at the site of a DNA repair, epigenetic changes can remain. In particular, a double strand break in DNA can initiate unprogrammed epigenetic gene silencing both by causing DNA methylation as well as by promoting silencing types of histone modifications. In addition, the enzyme Parp1 (poly(ADP)-ribose polymerase) and its product poly(ADP)-ribose (PAR) accumulate at sites of DNA damage as part of a repair process. This accumulation, in turn, directs recruitment and activation of the chromatin remodeling protein ALC1 that can cause nucleosome remodeling. Nucleosome remodeling has been found to cause, for instance, epigenetic silencing of DNA repair gene MLH1. DNA damaging chemicals, such as benzene, hydroquinone, styrene, carbon tetrachloride and trichloroethylene, cause considerable hypomethylation of DNA, some through the activation of oxidative stress pathways.

Foods are known to alter the epigenetics of rats on different diets. Some food components epigenetically increase the levels of DNA repair enzymes such as MGMT and MLH1 and p53. Other food components can reduce DNA damage, such as soy isoflavones. In one study, markers for oxidative stress, such as modified nucleotides that can result from DNA damage, were decreased by a 3-week diet supplemented with soy. A decrease in oxidative DNA damage was also observed 2 h after consumption of anthocyanin-rich bilberry (*Vaccinium myrtillius* L.) pomace extract.

Techniques used to Study Epigenetics

Epigenetic research uses a wide range of molecular biological techniques to further understanding of epigenetic phenomena, including chromatin immunoprecipitation (together with its large-scale variants ChIP-on-chip and ChIP-Seq), fluorescent in situ hybridization, methylation-sensitive restriction enzymes, DNA adenine methyltransferase identification (DamID) and bisulfite sequencing. Furthermore, the use of bioinformatics methods has a role in (computational epigenetics).

Functions and Consequences

Development

Developmental epigenetics can be divided into predetermined and probabilistic epigenesis. Predetermined epigenesis is a unidirectional movement from structural development in DNA to the functional maturation of the protein. "Predetermined" here means that development is scripted and predictable. Probabilistic epigenesis on the other hand is a bidirectional structure-function development with experiences and external molding development.

Somatic epigenetic inheritance, particularly through DNA and histone covalent modifications and nucleosome repositioning, is very important in the development of multicellular eukaryotic organisms. The genome sequence is static (with some notable exceptions), but cells differentiate into many different types, which perform different functions, and respond differently to the environment and intercellular signalling. Thus, as individuals develop, morphogens activate or silence genes in an epigenetically heritable fashion, giving cells a memory. In mammals, most cells terminally differentiate, with only stem cells retaining the ability to differentiate into several cell types ("totipotency" and "multipotency"). In mammals, some stem cells continue producing new differentiated cells throughout life, such as in neurogenesis, but mammals are not able to respond to loss of some tissues, for example, the inability to regenerate limbs, which some other animals are capable of. Epigenetic modifications regulate the transition from neural stem cells to glial progenitor cells (for example, differentiation into oligodendrocytes is regulated by the deacetylation and methylation of histones. Unlike animals, plant cells do not terminally differentiate, remaining totipotent with the ability to give rise to a new individual plant. While plants do utilise many of the same epigenetic mechanisms as animals, such as chromatin remodeling, it has been hypothesised that some kinds of plant cells do not use or require "cellular memories", resetting their gene expression patterns using positional information from the environment and surrounding cells to determine their fate.

Epigenetic changes can occur in response to environmental exposure – for example, mice given some dietary supplements have epigenetic changes affecting expression of the agouti gene, which affects their fur color, weight, and propensity to develop cancer.

Controversial results from one study suggested that traumatic experiences might produce an epigenetic signal that is capable of being passed to future generations. Mice were trained, using foot shocks, to fear a cherry blossom odor. The investigators reported that the mouse offspring had an increased aversion to this specific odor. They suggested epigenetic changes that increase gene expression, rather than in DNA itself, in a gene, M71, that governs the functioning of an odor receptor in the nose that responds specifically to this cherry blossom smell. There were physical changes that correlated with olfactory (smell) function in the brains of the trained mice and their descendants. Several criticisms were reported, including the study's low statistical power as evidence of some irregularity such as bias in reporting results. Due to limits of sample size, there is a probability that an effect will not be demonstrated to within statistical significance even if it exists. The criticism suggested that the probability that all the experiments reported would show positive results if an identical protocol was followed, assuming the claimed effects exist, is merely 0.4%. The authors also did not indicate which mice were siblings, and treated all of the mice as statistically independent. The original researchers pointed out negative results in the paper's appendix that the criticism omitted in its calculations, and undertook to track which mice were siblings in the future.

Transgenerational

Epigenetic mechanisms were a necessary part of the evolutionary origin of cell differentiation.

Although epigenetics in multicellular organisms is generally thought to be a mechanism involved in differentiation, with epigenetic patterns "reset" when organisms reproduce, there have been some observations of transgenerational epigenetic inheritance (e.g., the phenomenon of paramutation observed in maize). Although most of these multigenerational epigenetic traits are gradually lost over several generations, the possibility remains that multigenerational epigenetics could be another aspect to evolution and adaptation. As mentioned above, some define epigenetics as heritable.

A sequestered germ line or Weismann barrier is specific to animals, and epigenetic inheritance is more common in plants and microbes. Eva Jablonka, Marion J. Lamb and Étienne Danchin have argued that these effects may require enhancements to the standard conceptual framework of the modern synthesis and have called for an extended evolutionary synthesis. Other evolutionary biologists, such as John Maynard Smith, have incorporated epigenetic inheritance into population genetics models or are openly skeptical of the extended evolutionary synthesis (Michael Lynch). Thomas Dickins and Qazi Rahman state that epigenetic mechanisms such as DNA methylation and histone modification are genetically inherited under the control of natural selection and therefore fit under the earlier "modern synthesis".

Two important ways in which epigenetic inheritance can be different from traditional genetic inheritance, with important consequences for evolution, are that rates of epimutation can be much faster than rates of mutation and the epimutations are more easily reversible. In plants, heritable DNA methylation mutations are 100,000 times more likely to occur compared to DNA mutations. An epigenetically inherited element such as the PSI+ system can act as a "stop-gap", good enough for short-term adaptation that allows the lineage to survive for long enough for mutation and/or recombination to genetically assimilate the adaptive phenotypic change. The existence of this possibility increases the evolvability of a species.

More than 100 cases of transgenerational epigenetic inheritance phenomena have been reported in a wide range of organisms, including prokaryotes, plants, and animals. For instance, mourning cloak butterflies will change color through hormone changes in response to experimentation of varying temperatures.

The filamentous fungus *Neurospora crassa* is a prominent model system for understanding the control and function of cytosine methylation. In this organism, DNA methylation is associated with relics of a genome defense system called RIP (repeat-induced point mutation) and silences gene expression by inhibiting transcription elongation.

The yeast prion PSI is generated by a conformational change of a translation termination factor, which is then inherited by daughter cells. This can provide a survival advantage under adverse conditions. This is an example of epigenetic regulation enabling unicellular organisms to respond rapidly to environmental stress. Prions can be viewed as epigenetic agents capable of inducing a phenotypic change without modification of the genome.

Direct detection of epigenetic marks in microorganisms is possible with single molecule real time sequencing, in which polymerase sensitivity allows for measuring methylation and other modifications as a DNA molecule is being sequenced. Several projects have demonstrated the ability to collect genome-wide epigenetic data in bacteria.

Epigenetics in Bacteria

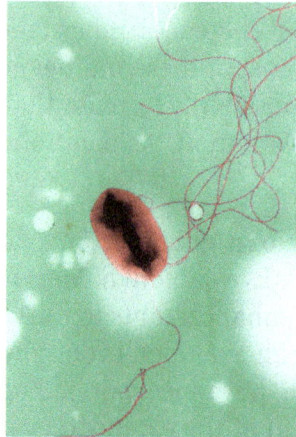

Escherichia coli bacteria.

While epigenetics is of fundamental importance in eukaryotes, especially metazoans, it plays a different role in bacteria. Most importantly, eukaryotes use epigenetic mechanisms primarily to regulate gene expression which bacteria rarely do. However, bacteria make widespread use of post-replicative DNA methylation for the epigenetic control of DNA-protein interactions. Bacteria also use DNA adenine methylation (rather than DNA cytosine methylation) as an epigenetic signal. DNA adenine methylation is important in bacteria virulence in organisms such as Escherichia coli, Salmonella, Vibrio, Yersinia, Haemophilus, and Brucella. In Alphaproteobacteria, methylation of adenine regulates the cell cycle and couples gene transcription to DNA replication. In Gammaproteobacteria, adenine methylation provides signals for DNA replication, chromosome segregation, mismatch repair, packaging of bacteriophage, transposase activity and regulation of gene expression. There exists a genetic switch controlling Streptococcus pneumoniae (the pneumococcus) that allows the bacterium to randomly change its characteristics into six alternative states that could pave the way to improved vaccines. Each form is randomly generated by a phase variable methylation system. The ability of the pneumococcus to cause deadly infections is different in each of these six states. Similar systems exist in other bacterial genera.

Medicine

Epigenetics has many and varied potential medical applications. In 2008, the National Institutes of Health announced that $190 million had been earmarked for epigenetics research over the next five years. In announcing the funding, government officials noted that epigenetics has the potential to explain mechanisms of aging, human development, and the origins of cancer, heart disease, mental illness, as well as several other conditions. Some investigators, like Randy Jirtle, PhD, of Duke University Medical Center, think epigenetics may ultimately turn out to have a greater role in disease than genetics.

Twins

Direct comparisons of identical twins constitute an optimal model for interrogating environmental epigenetics. In the case of humans with different environmental exposures, monozygotic (identical) twins were epigenetically indistinguishable during their early years, while older twins had remarkable differences in the overall content and genomic distribution of 5-methylcytosine DNA and histone acetylation. The twin pairs who had spent less of their lifetime together and/or had greater differences in their medical histories were those who showed the largest differences in their levels of 5-methylcytosine DNA and acetylation of histones H3 and H4.

Dizygotic (fraternal) and monozygotic (identical) twins show evidence of epigenetic influence in humans. DNA sequence differences that would be abundant in a singleton-based study do not interfere with the analysis. Environmental differences can produce long-term epigenetic effects, and different developmental monozygotic twin subtypes may be different with respect to their susceptibility to be discordant from an epigenetic point of view.

A high-throughput study, which denotes technology that looks at extensive genetic markers, focused on epigenetic differences between monozygotic twins to compare global and locus-specific changes in DNA methylation and histone modifications in a sample of 40 monozygotic twin pairs. In this case, only healthy twin pairs were studied, but a wide range of ages was represented, between 3 and 74 years. One of the major conclusions from this study was that there is an age-dependent accumulation of epigenetic differences between the two siblings of twin pairs. This accumulation suggests the existence of epigenetic "drift". Epigenetic drift is the term given to epigenetic modifications as they occur as a direct function with age. While age is a known risk factor for many diseases, age-related methylation has been found to occur differentially at specific sites along the genome. Over time, this can result in measurable differences between biological and chronological age. Epigenetic changes have been found to be reflective of lifestyle and may act as functional biomarkers of disease before clinical threshold is reached.

A more recent study, where 114 monozygotic twins and 80 dizygotic twins were analyzed for the DNA methylation status of around 6000 unique genomic regions, concluded that epigenetic similarity at the time of blastocyst splitting may also contribute to phenotypic similarities in monozygotic co-twins. This supports the notion that microenvironment at early stages of embryonic development can be quite important for the establishment of epigenetic marks. Congenital genetic disease is well understood and it is clear that epigenetics can play a role, for example, in the case of Angelman syndrome and Prader-Willi syndrome. These are normal genetic diseases caused by gene deletions or inactivation of the genes, but are unusually common because individuals are essentially hemizygous because of genomic imprinting, and therefore a single gene knock out is sufficient to cause the disease, where most cases would require both copies to be knocked out.

Genomic Imprinting

Some human disorders are associated with genomic imprinting, a phenomenon in mammals where the father and mother contribute different epigenetic patterns for specific genomic loci in their germ cells. The best-known case of imprinting in human disorders is that of Angelman syndrome and Prader-Willi syndrome – both can be produced by the same genetic mutation, chromosome 15q partial deletion, and the particular syndrome that will develop depends on whether the mutation is inherited from the child's mother or from their father. This is due to the presence of genomic imprinting in the region. Beckwith-Wiedemann syndrome is also associated with genomic imprinting, often caused by abnormalities in maternal genomic imprinting of a region on chromosome 11.

Rett syndrome is underlain by mutations in the MECP2 gene despite no large-scale changes in expression of MeCP2 being found in microarray analyses. BDNF is downregulated in the MECP2 mutant resulting in Rett syndrome.

In the Överkalix study, paternal (but not maternal) grandsons of Swedish men who were exposed during preadolescence to famine in the 19th century were less likely to die of cardiovascular disease. If food was plentiful, then diabetes mortality in the grandchildren increased, suggesting that this was a transgenerational epigenetic inheritance. The opposite effect was observed for females – the paternal (but not maternal) granddaughters of women who experienced famine while in the womb (and therefore while their eggs were being formed) lived shorter lives on average.

Cancer

A variety of epigenetic mechanisms can be perturbed in different types of cancer. Epigenetic alterations of DNA repair genes or cell cycle control genes are very frequent in sporadic (non-germ line) cancers, being significantly more common than germ line (familial) mutations in these sporadic cancers. Epigenetic alterations are important in cellular transformation to cancer, and their manipulation holds great promise for cancer prevention, detection, and therapy. Several medications which have epigenetic impact are used in several of these diseases. These aspects of epigenetics are addressed in cancer epigenetics.

Diabetic Wound Healing

Epigenetic modifications have given insight into the understanding of the pathophysiology of different disease conditions. Though, they are strongly associated with cancer, their role in other pathological conditions are of equal importance. It appears that, the hyperglycaemic environment could imprint such changes at the genomic level, that macrophages are primed towards a pro-inflammatory state and could fail to exhibit any phenotypic alteration towards the pro-healing type. This phenomenon of altered Macrophage Polarization is mostly associated with all the diabetic complications in a clinical set-up. At present, several reports reveal the relevance of different epigenetic modifications with respect to diabetic complications. Sooner or later, with the advancements in biomedical tools, the detection of such biomarkers as prognostic and diagnostic tools in patients could possibly emerge out as alternative approaches. It is noteworthy to mention here that the use of epigenetic modifications as therapeutic targets warrant extensive preclinical as well as clinical evaluation prior to use.

Psychology and Psychiatry

Early Life Stress

In a groundbreaking 2003 report, Caspi and colleagues demonstrated that in a robust cohort of over one-thousand subjects assessed multiple times from preschool to adulthood, subjects who carried one or two copies of the short allele of the serotonin transporter promoter polymorphism exhibited higher rates of adult depression and suicidality when exposed to childhood maltreatment when compared to long allele homozygotes with equal ELS exposure.

Parental nutrition, in utero exposure to stress, male-induced maternal effects such as attraction of differential mate quality, and maternal as well as paternal age, and offspring gender could all possibly influence whether a germline epimutation is ultimately expressed in offspring and the degree to which intergenerational inheritance remains stable throughout posterity.

Addiction

Addiction is a disorder of the brain's reward system which arises through transcriptional and neuroepigenetic mechanisms and occurs over time from chronically high levels of exposure to an addictive stimulus (e.g., morphine, cocaine, sexual intercourse, gambling, etc.). Transgenerational epigenetic inheritance of addictive phenotypes has been noted to occur in preclinical studies.

Anxiety

Transgenerational epigenetic inheritance of anxiety-related phenotypes has been reported in a preclinical study using mice. In this investigation, transmission of paternal stress-induced traits across generations involved small non-coding RNA signals transmitted via the male germline.

Depression

Epigenetic inheritance of depression-related phenotypes has also been reported in a preclinical study. Inheritance of paternal stress-induced traits across generations involved small non-coding RNA signals transmitted via the paternal germline.

Fear Conditioning

Studies on mice have shown that certain conditional fears can be inherited from either parent. In one example, mice were conditioned to fear a strong scent, acetophenone, by accompanying the smell with an electric shock. Consequently, the mice learned to fear the scent of acetophenone alone. It was discovered that this fear could be passed down to the mice offspring. Despite the offspring never experiencing the electric shock themselves the mice still displayed a fear of the acetophenone scent, because they inherited the fear epigenetically by site-specific DNA methylation. These epigenetic changes lasted up to two generations without reintroducing the shock.

Mechanism of Epigenetics

Epigenetics acts mainly through four different mechanisms.

- DNA Methylation.

- Chromatin remodelling.

- Histone modification.

- RNA interference/interactions.

Modification at the DNA Level

These modifications are cancelled during the process of gametogenesis and embryogenesis. Thus it is an epigenetic phenomenon but not inherited over the generations.

- Cytosine methylation: It is the addition of methyl group to the Cytosine base of the DNA sequence to form 5' methyl Cytosine (Thymine). Any mutation at 5' methyl cytosine site in the DNA sequence converts it into Uracil and therefore hard to be identified and repaired.

- Methylation at promoter site: When the methylation occurs at the promoter site, the transcription process gets suppressed.

Methylation is important for:

- Silencing transcription.

- Genomic imprinting.

- X chromosome inactivation.

- Protecting the genome from transposition.

- Tissue specific gene expression and regulation.

- Heterochromatin.

- Developmental controls.

- Cancer therapy.

Defects in the methylation cause diseases such as Systemic lupus erythematosus (SLE), Immunodeficiency, and facial anomalies (ICF) syndrome. Demethylation of promoter sequences increases chemosensitivity, adhesion, response to interferons, and immunogenicity while it helps to decrease growth of cancerous cells.

Chromatin Remodeling

Chromatin remodeling includes the shifting of nucleosome cores. The process is known as nucleosome sliding. The shift results from disassembling and reassembling the units of nucleosome core. The process is one of the major factors controlling the gene expression by induction and repression.

The remodeling is brought about by SWI/SNF family of ATPase complexes. There are four types of complexes based on the type of ATPases. These are SWI2/SNF2, imitation switch (ISWI); INO80; and Mi-2 (CHD1).

Histone Modification

There are a lot of histone modifications which are typically conserved over evolutionary processes. These include:

- Methylation of lysine and arginine residues.

- Acetylaton of lysine.

- Phosphorylation of serine and threonine residues.

- Proline isomerization.

- Monoubiquitylation.

- Sumoylation.

These modifications can occur in both coding as well as non coding sequences of genome. Some modifications are specific to either active or inactive regions of transcription. These are influenced by the developmental stage, phase in cell cycle, stress, and other environmental factors.

- Histone acetylation: Acetylated histones are found to open the chromatin and thus enable transcription. The acetylation of Histones are carried out by histone acetylases. These enzymes form part of chromatin remodeling and transcription complexes. The N-terminal Lysine residues undergo acetylation and this causes the Histone proteins to lose their positive charges. The affinity between DNA and histones gets reduces and the promoter regions become easily accessible to the enzymes initiating transcription.

  When histones are deacetylated, they are less aceessible to transcription and are tightly packed. The deacetylation of histones is carried out by the enzyme HDAC or histone deacylase.

- Histone methylation.

- Histone phosphorylation: Phosphorylation of Histones occurs during mitosis, signal transduction pathways like the ERK pathway.

- Histone ubiquitilation: Ubiquitilation of histones has been found to cause heritable gene silencing and inactivatin of X chromosome.

RNA Interference

Epigenetic regulation is influenced by the presence of non protein coding RNAs. These RNAs form an essential part in RNA interference. The resultant small double stranded RNA molecules inhibit gene expression by interacting with the nascent RNA molecule, DNA sequence, or by other mechanisms involving chromatin modifiers. These siRNA molecules are involved in the formation of RISCs (RNA induced Silencing Complexes). The complexes thus formed promote epigenetic silencing by cleavage of RNA molecule or through RNA directed DNA methylation. The process of translation gets inhibited temporarily but doesn't eliminate the gene expression.

References

- Gene, entry: newworldencyclopedia.org, Retrieved 14 July, 2019

- Sangeeta Jain; Nima Goharkhay; George Saade; Gary D. Hankins; Garland D. Anderson (2007). "Hepatitis C in pregnancy". American Journal of Perinatology. 24 (4): 251–256. Doi:10.1055/s-2007-970181

- Genetics: biologydictionary.net, Retrieved 17 May, 2019

- Bird A (January 2002). "DNA methylation patterns and epigenetic memory". Genes & Development. 16 (1): 6–21. Doi:10.1101/gad.947102. PMID 11782440

- Culver KW, Labow MA (8 November 2002). "Genomics". In Robinson R (ed.). Genetics. Macmillan Science Library. Macmillan Reference USA. ISBN 978-0-02-865606-9

- Lobo, Ingrid; Shaw, Kenna. "Discovery and Types of Genetic Linkage". Scitable. Nature Education. Retrieved 21 January 2017

- Gallvetti, Andrea; Whipple, Clinton J. (2015). "Positional cloning in maize (Zea mays subsp. Mays, Poaceae)". Applications in Plant Sciences. 3 (1): 1400092. Doi:10.3732/apps.1400092. PMC 4298233. PMID 25606355

4
Molecular Biology

The branch of biology which focuses on the molecular basis of biological activity between bio-molecules in different systems of a cell is called molecular biology. Some of the diverse techniques used within this field are polymerase chain reactions, southern blotting and northern blotting. This chapter closely examines these key techniques of molecular biology to provide an extensive understanding of the subject.

Molecular biology is the field of science concerned with studying the chemical structures and processes of biological phenomena that involve the basic units of life, molecules. Of growing importance since the 1940s, molecular biology developed out of the related fields of biochemistry, genetics, and biophysics. The discipline is particularly concerned with the study of proteins and nucleic acids—i.e., the macromolecules that are essential to life processes. Molecular biology seeks to understand the three-dimensional structure of these macromolecules through such techniques as X-ray diffraction and electron microscopy. The discipline particularly seeks to understand the molecular basis of genetic processes; molecular biologists map the location of genes on specific chromosomes, associate these genes with particular characters of an organism, and use recombinant DNA technology to isolate, sequence, and modify specific genes.

In its early period during the 1940s, the field was concerned with elucidating the basic three-dimensional structure of proteins. Growing knowledge of the structure of proteins in the early 1950s enabled the structure of deoxyribonucleic acid (DNA)—the genetic blueprint found in all living things—to be described in 1953. Further research enabled scientists to gain an increasingly detailed knowledge not only of DNA and ribonucleic acid (RNA) but also of the chemical sequences within these substances that instruct the cells and viruses to make proteins.

Molecular biology remained a pure science with few practical applications until the 1970s, when certain types of enzymes were discovered that could cut and recombine segments of DNA in the chromosomes of certain bacteria. The resulting recombinant DNA technology became one of the most active branches of molecular biology because it allows the manipulation of the genetic sequences that determine the basic characters of organisms.

Molecular Genetics

Molecular genetics is the field of biology that studies the structure and function of genes at a molecular level and thus employs methods of both molecular biology and genetics. The study of chromosomes and gene expression of an organism can give insight into heredity, genetic variation, and mutations. This is useful in the study of developmental biology and in understanding and treating genetic diseases.

Technique

Amplification

Gene amplification is a procedure in which a certain gene or DNA sequence is replicated many times in a process called DNA replication.

Polymerase Chain Reaction

The main genetic components of the polymerase chain reaction (PCR) are DNA nucleotides, template DNA, primers and Taq polymerase. DNA nucleotides make up the DNA template strand for the specific sequence being amplified and primers are short strands of complementary nucleotides where DNA replication starts. Taq polymerase is a heat stable enzyme that jump-starts the production of new DNA at the high temperatures needed for reaction.

Cloning DNA in Bacteria

Cloning is the process of creating many identical copies of a sequence of DNA. The target DNA sequence is inserted into a cloning vector. Because this vector originates from a self-replicating virus, plasmid, or higher organism cell, when the appropriate size DNA is inserted, the "target and vector DNA fragments are then ligated" to create a recombinant DNA molecule.

The recombinant DNA molecule is then inserted into a bacterial strain (usually *E. coli*) which produces several identical copies of the selected sequence it absorbed through transformation (the mechanism by which bacteria uptake foreign DNA from the environment into their genomes).

Separation and Detection

In separation and detection, DNA and mRNA are isolated from cells and then detected simply by the isolation. Cell cultures are also grown to provide a constant supply of cells ready for isolation.

Cell Cultures

A cell culture for molecular genetics is a culture that is grown in artificial conditions. Some cell types grow well in cultures such as skin cells, but other cells are not as productive in cultures. There are different techniques for each type of cell, some only recently being found to foster growth in stem and nerve cells. Cultures for molecular genetics are frozen in order to preserve all copies of the gene specimen and thawed only when needed. This allows for a steady supply of cells.

DNA Isolation

DNA isolation extracts DNA from a cell in a pure form. First, the DNA is separated from cellular components such as proteins, RNA, and lipids. This is done by placing the chosen cells in a tube with a solution that mechanically, chemically, breaks the cells open. This solution contains enzymes, chemicals, and salts that breaks down the cells except for the DNA. It contains enzymes to dissolve proteins, chemicals to destroy all RNA present, and salts to help pull DNA out of the solution. Next, the DNA is separated from the solution by being spun in a centrifuge, which allows the DNA to collect in the bottom of the tube. After this cycle in the centrifuge the solution is poured off and the DNA is resuspended in a second solution that makes the DNA easy to work with in the future. This results in a concentrated DNA sample that contains thousands of copies of each gene. For large scale projects such as sequencing the human genome, all this work is done by robots.

mRNA Isolation

Expressed DNA that codes for the synthesis of a protein is the final goal for scientists and this expressed DNA is obtained by isolating mRNA (Messenger RNA).

First, laboratories use a normal cellular modification of mRNA that adds up to 200 adenine nucleotides to the end of the molecule (poly(A) tail). Once this has been added, the cell is ruptured and its cell contents are exposed to synthetic beads that are coated with thymine string nucleotides. Because Adenine and Thymine pair together in DNA, the poly(A) tail and synthetic beads are attracted to one another, and once they bind in this process the cell components can be washed away without removing the mRNA. Once the mRNA has been isolated, reverse transcriptase is employed to convert it to single-stranded DNA, from which a stable double-stranded DNA is produced using DNA polymerase. Complementary DNA (cDNA) is much more stable than mRNA and so, once the double-stranded DNA has been produced it represents the expressed DNA sequence scientists look for.

Genetic Screens

Forward Genetics

This technique is used to identify which genes or genetic mutations produce a certain phenotype. A mutagen is very often used to accelerate this process. Once mutants have been isolated, the mutated genes can be molecularly identified.

Forward saturation genetics is a method for treating organisms with a mutagen, then screens the organism's offspring for particular phenotypes. This type of genetic screening is used to find and identify all the genes involved in a trait.

Reverse Genetics

Reverse genetics determines the phenotype that results from a specifically engineered gene. In some organisms, such as yeast and mice, it is possible to induce the deletion of a particular gene, creating what's known as a gene "knockout" - the laboratory origin of so-called "knockout mice" for further study. In other words this process involves the creation of transgenic organisms that do not express a gene of interest. Alternative methods of reverse genetic research include the random induction of DNA deletions and subsequent selection for deletions in a gene of interest, as well as the application of RNA interference.

Gene Therapy

A mutation in a gene can cause encoded proteins and the cells that rely on those proteins to malfunction. Conditions related to gene mutations are called genetic disorders. However, altering a patient's genes can sometimes be used to treat or cure a disease as well. Gene therapy can be used to replace a mutated gene with the correct copy of the gene, to inactivate or knockout the expression of a malfunctioning gene, or to introduce a foreign gene to the body to help fight disease. Major diseases that can be treated with gene therapy include viral infections, cancers, and inherited disorders, including immune system disorders.

Gene therapy delivers a copy of the missing, mutated, or desired gene via a modified virus or vector to the patient's target cells so that a functional form of the protein can then be produced and incorporated into the body. These vectors are often siRNA. Treatment can be either in vivo or ex vivo. The therapy has to be repeated several times for the infected patient to continually be relieved, as repeated cell division and cell death slowly reduces the body's ratio of functional-to-mutant genes. Gene therapy is an appealing alternative to some drug-based approaches, because gene therapy repairs the underlying genetic defect using the patients own cells with minimal side effects. Gene therapies are still in development and mostly used in research settings. All experiments and products are controlled by the U.S. FDA and the NIH.

Classical gene therapies usually require efficient transfer of cloned genes into the disease cells so that the introduced genes are expressed at sufficiently high levels to change the patient's physiology. There are several different physicochemical and biological methods that can be used to transfer genes into human cells. The size of the DNA fragments that can be transferred is very limited, and often the transferred gene is not a conventional gene. Horizontal gene transfer is the transfer of genetic material from one cell to another that is not its offspring. Artificial horizontal gene transfer is a form of genetic engineering.

The Human Genome Project

The Human Genome Project is a molecular genetics project that began in the 1990s and was projected to take fifteen years to complete. However, because of technological advances the progress of the project was advanced and the project finished in 2003, taking only thirteen years. The project was started by the U.S. Department of Energy and the National Institutes of Health in an effort to reach six set goals. These goals included:

- Identifying 20,000 to 25,000 genes in human dna (although initial estimates were approximately 100,000 genes),

- Determining sequences of chemical base pairs in human dna,

- Storing all found information into databases,

- Improving the tools used for data analysis,

- Transferring technologies to private sectors, and

- Addressing the ethical, legal, and social issues (ELSI) that may arise from the projects.

The project was worked on by eighteen different countries including the United States, Japan, France, Germany, and the United Kingdom. The collaborative effort resulted in the discovery of the many benefits of molecular genetics. Discoveries such as molecular medicine, new energy sources and environmental applications, DNA forensics, and livestock breeding, are only a few of the benefits that molecular genetics can provide.

Molecular Cloning

Molecular cloning is a set of experimental methods in molecular biology that are used to assemble recombinant DNA molecules and to direct their replication within host organisms. The use of the word cloning refers to the fact that the method involves the replication of one molecule to produce a population of cells with identical DNA molecules. Molecular cloning generally uses DNA sequences from two different organisms: the species that is the source of the DNA to be cloned, and the species that will serve as the living host for replication of the recombinant DNA. Molecular cloning methods are central to many contemporary areas of modern biology and medicine.

In a conventional molecular cloning experiment, the DNA to be cloned is obtained from an organism of interest, then treated with enzymes in the test tube to generate smaller DNA fragments. Subsequently, these fragments are then combined with vector DNA to generate recombinant DNA molecules. The recombinant DNA is then introduced into a host organism (typically an easy-to-grow, benign, laboratory strain of E. coli bacteria). This will generate a population of organisms in which recombinant DNA molecules are replicated along with the host DNA. Because they contain foreign DNA fragments, these are transgenic or genetically modified microorganisms (GMO). This process takes advantage of the fact that a single bacterial cell can be induced to take up and replicate a single recombinant DNA molecule. This single cell can then be expanded exponentially to generate a large amount of bacteria, each of which contain copies of the original recombinant molecule. Thus, both the resulting bacterial population, and the recombinant DNA molecule, are commonly referred to as "clones". Strictly speaking, recombinant DNA refers to DNA molecules, while molecular cloning refers to the experimental methods used to assemble them. The idea arose that different DNA sequences could be inserted into a plasmid and that these foreign sequences would be carried into bacteria and digested as part of the plasmid. That is, these plasmids could serve as cloning vectors to carry genes.

Virtually any DNA sequence can be cloned and amplified, but there are some factors that might limit the success of the process. Examples of the DNA sequences that are difficult to clone are inverted repeats, origins of replication, centromeres and telomeres. Another characteristic that limits chances of success is large size of DNA sequence. Inserts larger than 10kbp have very limited success, but bacteriophages such as bacteriophage λ can be modified to successfully insert a sequence up to 40 kbp.

Prior to the 1970s, the understanding of genetics and molecular biology was severely hampered by an inability to isolate and study individual genes from complex organisms. This changed dramatically with the advent of molecular cloning methods. Microbiologists, seeking to understand the molecular mechanisms through which bacteria restricted the growth of bacteriophage, isolated restriction endonucleases, enzymes that could cleave DNA molecules only when specific DNA sequences were encountered. They showed that restriction enzymes cleaved chromosome-length DNA molecules at specific locations, and that specific sections of the larger molecule could be purified by size fractionation. Using a second enzyme, DNA ligase, fragments generated by restriction enzymes could be joined in new combinations, termed recombinant DNA. By recombining DNA segments of interest with vector DNA, such as bacteriophage or plasmids, which naturally replicate inside bacteria, large quantities of purified recombinant DNA molecules could be produced in bacterial cultures. The first recombinant DNA molecules were generated and studied in 1972.

Molecular cloning takes advantage of the fact that the chemical structure of DNA is fundamentally the same in all living organisms. Therefore, if any segment of DNA from any organism is inserted into a DNA segment containing the molecular sequences required for DNA replication, and the resulting recombinant DNA is introduced into the organism from which the replication sequences were obtained, then the foreign DNA will be replicated along with the host cell's DNA in the transgenic organism.

Molecular cloning is similar to polymerase chain reaction (PCR) in that it permits the replication of DNA sequence. The fundamental difference between the two methods is that molecular cloning involves replication of the DNA in a living microorganism, while PCR replicates DNA in an *in vitro* solution, free of living cells.

Steps

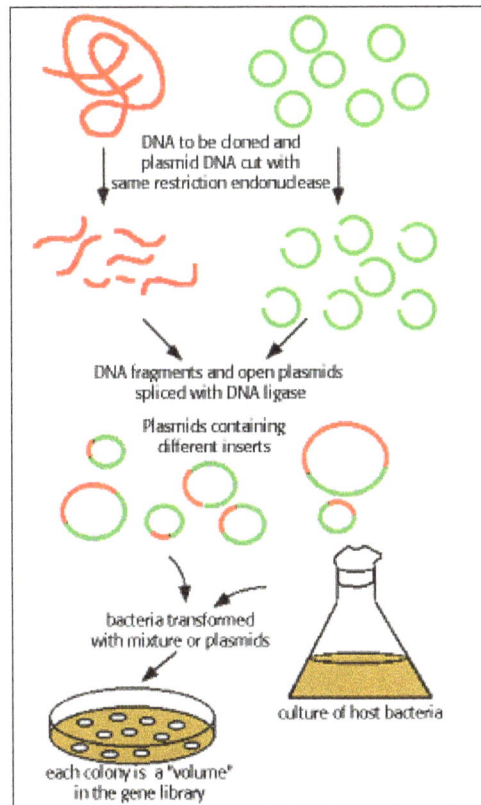

The overall goal of molecular cloning is to take a gene of interest from one plasmid and insert it into another plasmid.

In standard molecular cloning experiments, the cloning of any DNA fragment essentially involves seven steps: (1) Choice of host organism and cloning vector, (2) Preparation of vector DNA, (3) Preparation of DNA to be cloned, (4) Creation of recombinant DNA, (5) Introduction of recombinant DNA into host organism, (6) Selection of organisms containing recombinant DNA, (7) Screening for clones with desired DNA inserts and biological properties.

Although the detailed planning of the cloning can be done in any text editor, together with online utilities for e.g. PCR primer design, dedicated software exist for the purpose. Software for the purpose include for example ApE (open source), DNAStrider (open source), Serial Cloner (gratis) and Collagene (open source).

Notably, the growing capacity and fidelity of DNA synthesis platforms allows for increasingly intricate designs in molecular engineering. These projects may include very long strands of novel DNA sequence and/or test entire libraries simultaneously, as opposed to of individual sequences. These shifts introduce complexity that require design to move away from the flat nucleotide-based representation and towards a higher level of abstraction. Examples of such tools are GenoCAD, Teselagen (free for academia) or GeneticConstructor (free for academics).

Choice of Host Organism and Cloning Vector

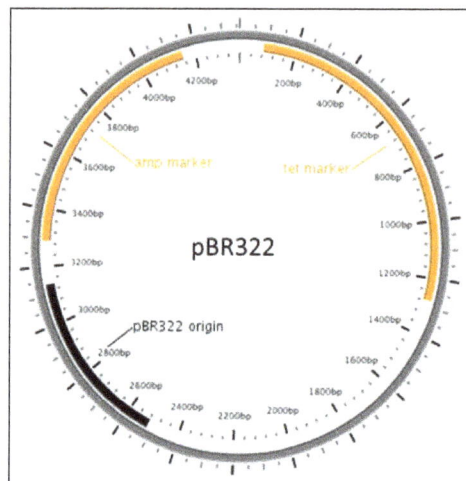

Diagram of a commonly used cloning plasmid; pBR322.

It's a circular piece of DNA 4361 bases long. Two antibiotic resistance genes are present, conferring resistance to ampicillin and tetracycline, and an origin of replication that the host uses to replicate the DNA.

Although a very large number of host organisms and molecular cloning vectors are in use, the great majority of molecular cloning experiments begin with a laboratory strain of the bacterium *E. coli* (*Escherichia coli*) and a plasmid cloning vector. *E. coli* and plasmid vectors are in common use because they are technically sophisticated, versatile, widely available, and offer rapid growth of recombinant organisms with minimal equipment. If the DNA to be cloned is exceptionally large (hundreds of thousands to millions of base pairs), then a bacterial artificial chromosome or yeast artificial chromosome vector is often chosen.

Specialized applications may call for specialized host-vector systems. For example, if the experimentalists wish to harvest a particular protein from the recombinant organism, then an expression vector is chosen that contains appropriate signals for transcription and translation in the desired host organism. Alternatively, if replication of the DNA in different species is desired (for example, transfer of DNA from bacteria to plants), then a multiple host range vector (also termed shuttle vector) may be selected. In practice, however, specialized molecular cloning experiments usually begin with cloning into a bacterial plasmid, followed by subcloning into a specialized vector.

Whatever combination of host and vector are used, the vector almost always contains four DNA segments that are critically important to its function and experimental utility:

- DNA replication origin is necessary for the vector (and its linked recombinant sequences) to replicate inside the host organism.

- One Or More unique restriction endonuclease recognition sites to serves as sites where foreign DNA may be introduced.

- A Selectable genetic marker gene that can be used to enable the survival of cells that have taken up vector sequences.

- A Tag gene that can be used to screen for cells containing the foreign DNA.

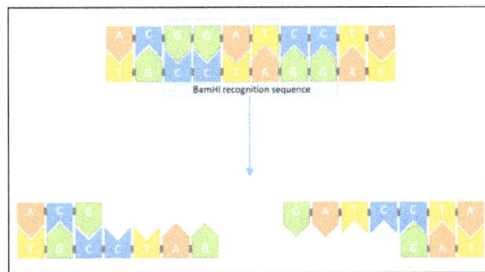

Cleavage of a DNA sequence containing the BamHI restriction site.
The DNA is cleaved at the palindromic sequence to produce 'sticky ends'.

Preparation of Vector DNA

The cloning vector is treated with a restriction endonuclease to cleave the DNA at the site where foreign DNA will be inserted. The restriction enzyme is chosen to generate a configuration at the cleavage site that is compatible with the ends of the foreign DNA. Typically, this is done by cleaving the vector DNA and foreign DNA with the same restriction enzyme, for example EcoRI. Most modern vectors contain a variety of convenient cleavage sites that are unique within the vector molecule (so that the vector can only be cleaved at a single site) and are located within a gene (frequently beta-galactosidase) whose inactivation can be used to distinguish recombinant from non-recombinant organisms at a later step in the process. To improve the ratio of recombinant to non-recombinant organisms, the cleaved vector may be treated with an enzyme (alkaline phosphatase) that dephosphorylates the vector ends. Vector molecules with dephosphorylated ends are unable to replicate, and replication can only be restored if foreign DNA is integrated into the cleavage site.

Preparation of DNA to be Cloned

Template DNA is mixed with bases (the building blocks of DNA), primers (short pieces of complementary single stranded DNA) and a DNA polymerase enzyme that builds the DNA chain. The mix goes through cycles of heating and cooling to produce large quantities of copied DNA.

For cloning of genomic DNA, the DNA to be cloned is extracted from the organism of interest. Virtually any tissue source can be used (even tissues from extinct animals), as long as the DNA is not extensively degraded. The DNA is then purified using simple methods to remove contaminating proteins (extraction with phenol), RNA (ribonuclease) and smaller molecules (precipitation and/

or chromatography). Polymerase chain reaction (PCR) methods are often used for amplification of specific DNA or RNA (RT-PCR) sequences prior to molecular cloning.

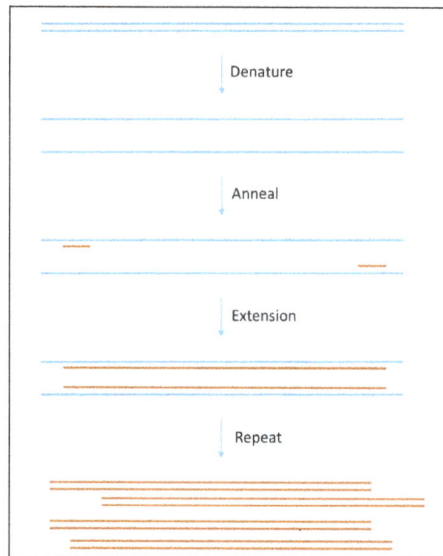

DNA for cloning is most commonly produced using PCR.

DNA for cloning experiments may also be obtained from RNA using reverse transcriptase (complementary DNA or cDNA cloning), or in the form of synthetic DNA (artificial gene synthesis). cDNA cloning is usually used to obtain clones representative of the mRNA population of the cells of interest, while synthetic DNA is used to obtain any precise sequence defined by the designer. Such a designed sequence may be required when moving genes across genetic codes (for example, from the mitochrondria to the nucleus) or simply for increasing expression via codon optimization.

The purified DNA is then treated with a restriction enzyme to generate fragments with ends capable of being linked to those of the vector. If necessary, short double-stranded segments of DNA (*linkers*) containing desired restriction sites may be added to create end structures that are compatible with the vector.

Creation of Recombinant DNA with DNA Ligase

The creation of recombinant DNA is in many ways the simplest step of the molecular cloning process. DNA prepared from the vector and foreign source are simply mixed together at appropriate concentrations and exposed to an enzyme (DNA ligase) that covalently links the ends together. This joining reaction is often termed ligation. The resulting DNA mixture containing randomly joined ends is then ready for introduction into the host organism.

DNA ligase only recognizes and acts on the ends of linear DNA molecules, usually resulting in a complex mixture of DNA molecules with randomly joined ends. The desired products (vector DNA covalently linked to foreign DNA) will be present, but other sequences (e.g. foreign DNA linked to itself, vector DNA linked to itself and higher-order combinations of vector and foreign DNA) are also usually present. This complex mixture is sorted out in subsequent steps of the cloning process, after the DNA mixture is introduced into cells.

Introduction of Recombinant DNA into Host Organism

The DNA mixture, previously manipulated in vitro, is moved back into a living cell, referred to as the host organism. The methods used to get DNA into cells are varied, and the name applied to this step in the molecular cloning process will often depend upon the experimental method that is chosen (e.g. transformation, transduction, transfection, electroporation).

When microorganisms are able to take up and replicate DNA from their local environment, the process is termed transformation, and cells that are in a physiological state such that they can take up DNA are said to be competent. In mammalian cell culture, the analogous process of introducing DNA into cells is commonly termed transfection. Both transformation and transfection usually require preparation of the cells through a special growth regime and chemical treatment process that will vary with the specific species and cell types that are used.

Electroporation uses high voltage electrical pulses to translocate DNA across the cell membrane (and cell wall, if present). In contrast, transduction involves the packaging of DNA into virus-derived particles, and using these virus-like particles to introduce the encapsulated DNA into the cell through a process resembling viral infection. Although electroporation and transduction are highly specialized methods, they may be the most efficient methods to move DNA into cells.

Selection of Organisms Containing Vector Sequences

Whichever method is used, the introduction of recombinant DNA into the chosen host organism is usually a low efficiency process; that is, only a small fraction of the cells will actually take up DNA. Experimental scientists deal with this issue through a step of artificial genetic selection, in which cells that have not taken up DNA are selectively killed, and only those cells that can actively replicate DNA containing the selectable marker gene encoded by the vector are able to survive.

When bacterial cells are used as host organisms, the selectable marker is usually a gene that confers resistance to an antibiotic that would otherwise kill the cells, typically ampicillin. Cells harboring the plasmid will survive when exposed to the antibiotic, while those that have failed to take up plasmid sequences will die. When mammalian cells (e.g. human or mouse cells) are used, a similar strategy is used, except that the marker gene (in this case typically encoded as part of the kanMX cassette) confers resistance to the antibiotic Geneticin.

Screening for Clones with Desired DNA Inserts and Biological Properties

Modern bacterial cloning vectors (e.g. pUC19 and later derivatives including the pGEM vectors) use the blue-white screening system to distinguish colonies (clones) of transgenic cells from those that contain the parental vector (i.e. vector DNA with no recombinant sequence inserted). In these vectors, foreign DNA is inserted into a sequence that encodes an essential part of beta-galactosidase, an enzyme whose activity results in formation of a blue-colored colony on the culture medium that is used for this work. Insertion of the foreign DNA into the beta-galactosidase coding sequence disables the function of the enzyme, so that colonies containing transformed DNA remain colorless (white). Therefore, experimentalists are easily able to identify and conduct further studies on transgenic bacterial clones, while ignoring those that do not contain recombinant DNA.

The total population of individual clones obtained in a molecular cloning experiment is often termed a DNA library. Libraries may be highly complex (as when cloning complete genomic DNA from an organism) or relatively simple (as when moving a previously cloned DNA fragment into a different plasmid), but it is almost always necessary to examine a number of different clones to be sure that the desired DNA construct is obtained. This may be accomplished through a very wide range of experimental methods, including the use of nucleic acid hybridizations, antibody probes, polymerase chain reaction, restriction fragment analysis and/or DNA sequencing.

Applications

Molecular cloning provides scientists with an essentially unlimited quantity of any individual DNA segments derived from any genome. This material can be used for a wide range of purposes, including those in both basic and applied biological science. A few of the more important applications are summarized here.

Genome Organization and Gene Expression

Molecular cloning has led directly to the elucidation of the complete DNA sequence of the genomes of a very large number of species and to an exploration of genetic diversity within individual species, work that has been done mostly by determining the DNA sequence of large numbers of randomly cloned fragments of the genome, and assembling the overlapping sequences.

At the level of individual genes, molecular clones are used to generate probes that are used for examining how genes are expressed, and how that expression is related to other processes in biology, including the metabolic environment, extracellular signals, development, learning, senescence and cell death. Cloned genes can also provide tools to examine the biological function and importance of individual genes, by allowing investigators to inactivate the genes, or make more subtle mutations using regional mutagenesis or site-directed mutagenesis. Genes cloned into expression vectors for functional cloning provide a means to screen for genes on the basis of the expressed protein's function.

Production of Recombinant Proteins

Obtaining the molecular clone of a gene can lead to the development of organisms that produce the protein product of the cloned genes, termed a recombinant protein. In practice, it is frequently more difficult to develop an organism that produces an active form of the recombinant protein in desirable quantities than it is to clone the gene. This is because the molecular signals for gene expression are complex and variable, and because protein folding, stability and transport can be very challenging.

Many useful proteins are currently available as recombinant products. These include-(1) medically useful proteins whose administration can correct a defective or poorly expressed gene (e.g. recombinant factor VIII, a blood-clotting factor deficient in some forms of hemophilia, and recombinant insulin, used to treat some forms of diabetes), (2) proteins that can be administered to assist in a life-threatening emergency (e.g. tissue plasminogen activator, used to treat strokes), (3) recombinant subunit vaccines, in which a purified protein can be used to immunize patients against infectious diseases, without exposing them to the infectious agent itself (e.g. hepatitis B vaccine), and (4) recombinant proteins as standard material for diagnostic laboratory tests.

Transgenic Organisms

Once characterized and manipulated to provide signals for appropriate expression, cloned genes may be inserted into organisms, generating transgenic organisms, also termed genetically modified organisms (GMOs). Although most GMOs are generated for purposes of basic biological research, a number of GMOs have been developed for commercial use, ranging from animals and plants that produce pharmaceuticals or other compounds (pharming), herbicide-resistant crop plants, and fluorescent tropical fish (GloFish) for home entertainment.

Gene Therapy

Gene therapy involves supplying a functional gene to cells lacking that function, with the aim of correcting a genetic disorder or acquired disease. Gene therapy can be broadly divided into two categories. The first is alteration of germ cells, that is, sperm or eggs, which results in a permanent genetic change for the whole organism and subsequent generations. This "germ line gene therapy" is considered by many to be unethical in human beings. The second type of gene therapy, "somatic cell gene therapy", is analogous to an organ transplant. In this case, one or more specific tissues are targeted by direct treatment or by removal of the tissue, addition of the therapeutic gene or genes in the laboratory, and return of the treated cells to the patient. Clinical trials of somatic cell gene therapy began in the late 1990s, mostly for the treatment of cancers and blood, liver, and lung disorders.

Despite a great deal of publicity and promises, the history of human gene therapy has been characterized by relatively limited success. The effect of introducing a gene into cells often promotes only partial and/or transient relief from the symptoms of the disease being treated. Some gene therapy trial patients have suffered adverse consequences of the treatment itself, including deaths. In some cases, the adverse effects result from disruption of essential genes within the patient's genome by insertional inactivation. In others, viral vectors used for gene therapy have been contaminated with infectious virus. Nevertheless, gene therapy is still held to be a promising future area of medicine, and is an area where there is a significant level of research and development activity.

Polymerase Chain Reaction

Polymerase chain reaction (PCR) is a method widely used in molecular biology to make several copies of a specific DNA segment. Using PCR, copies of DNA sequences are exponentially amplified to generate thousands to millions of more copies of that particular DNA segment. PCR is now a common and often indispensable technique used in medical laboratory and clinical laboratory research for a broad variety of applications including biomedical research and criminal forensics. PCR was developed by Kary Mullis in 1983 while he was an employee of the Cetus Corporation. He was awarded the Nobel Prize in Chemistry in 1993 (along with Michael Smith) for his work in developing the method.

The vast majority of PCR methods rely on thermal cycling. Thermal cycling exposes reactants to repeated cycles of heating and cooling to permit different temperature-dependent reactions—specifically, DNA melting and enzyme-driven DNA replication. PCR employs two main reagents – primers (which are short single strand DNA fragments known as oligonucleotides that are a complementary sequence to the target DNA region) and a DNA polymerase. In the first step of

PCR, the two strands of the DNA double helix are physically separated at a high temperature in a process called DNA melting. In the second step, the temperature is lowered and the primers bind to the complementary sequences of DNA. The two DNA strands then become templates for DNA polymerase to enzymatically assemble a new DNA strand from free nucleotides, the building blocks of DNA. As PCR progresses, the DNA generated is itself used as a template for replication, setting in motion a chain reaction in which the original DNA template is exponentially amplified.

Almost all PCR applications employ a heat-stable DNA polymerase, such as Taq polymerase, an enzyme originally isolated from the thermophilic bacterium *Thermus aquaticus*. If the polymerase used was heat-susceptible, it would denature under the high temperatures of the denaturation step. Before the use of Taq polymerase, DNA polymerase had to be manually added every cycle, which was a tedious and costly process.

Applications of the technique include DNA cloning for sequencing, gene cloning and manipulation, gene mutagenesis; construction of DNA-based phylogenies, or functional analysis of genes; diagnosis and monitoring of hereditary diseases; amplification of ancient DNA; analysis of genetic fingerprints for DNA profiling (for example, in forensic science and parentage testing); and detection of pathogens in nucleic acid tests for the diagnosis of infectious diseases.

Placing a strip of eight PCR tubes into a thermal cycler.

Principles

PCR amplifies a specific region of a DNA strand (the DNA target). Most PCR methods amplify DNA fragments of between 0.1 and 10 kilo base pairs (kbp) in length, although some techniques allow for amplification of fragments up to 40 kbp. The amount of amplified product is determined by the available substrates in the reaction, which become limiting as the reaction progresses.

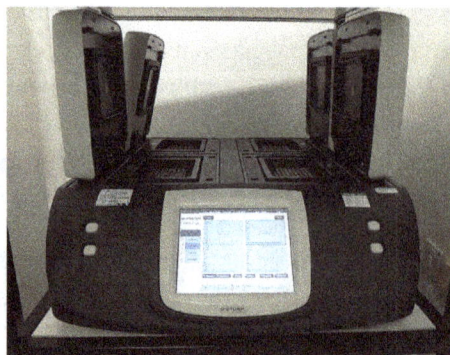

A thermal cycler for PCR.

A basic PCR set-up requires several components and reagents, including a DNA template that contains the DNA target region to amplify; a DNA polymerase; an enzyme that polymerizes new DNA strands; heat-resistant Taq polymerase is especially common, as it is more likely to remain intact during the high-temperature DNA denaturation process; two DNA primers that are complementary to the 3' (three prime) ends of each of the sense and anti-sense strands of the DNA target (DNA polymerase can only bind to and elongate from a double-stranded region of DNA; without primers there is no double-stranded initiation site at which the polymerase can bind); specific primers that are complementary to the DNA target region are selected beforehand, and are often custom-made in a laboratory or purchased from commercial biochemical suppliers; deoxynucleoside triphosphates, or dNTPs (sometimes called "deoxynucleotide triphosphates"; nucleotides containing triphosphate groups), the building blocks from which the DNA polymerase synthesizes a new DNA strand; a buffer solution providing a suitable chemical environment for optimum activity and stability of the DNA polymerase; bivalent cations, typically magnesium (Mg) or manganese (Mn) ions; Mg_2+ is the most common, but Mn_2+ can be used for PCR-mediated DNA mutagenesis, as a higher Mn_2+ concentration increases the error rate during DNA synthesis; and monovalent cations, typically potassium (K) ions.

An older, three-temperature thermal cycler for PCR.

The reaction is commonly carried out in a volume of 10–200 µL in small reaction tubes (0.2–0.5 mL volumes) in a thermal cycler. The thermal cycler heats and cools the reaction tubes to achieve the temperatures required at each step of the reaction. Many modern thermal cyclers make use of the Peltier effect, which permits both heating and cooling of the block holding the PCR tubes simply by reversing the electric current. Thin-walled reaction tubes permit favorable thermal conductivity to allow for rapid thermal equilibration. Most thermal cyclers have heated lids to prevent condensation at the top of the reaction tube. Older thermal cyclers lacking a heated lid require a layer of oil on top of the reaction mixture or a ball of wax inside the tube.

Procedure

Typically, PCR consists of a series of 20–40 repeated temperature changes, called thermal cycles, with each cycle commonly consisting of two or three discrete temperature steps. The cycling is often preceded by a single temperature step at a very high temperature (>90 °C (194 °F)), and followed by one hold at the end for final product extension or brief storage. The temperatures used and the length of time they are applied in each cycle depend on a variety of parameters, including the enzyme used for DNA synthesis, the concentration of bivalent ions and dNTPs in the reaction, and the melting temperature (T_m) of the primers. The individual steps common to most PCR methods are as follows:

- Initialization: This step is only required for DNA polymerases that require heat activation by hot-start PCR. It consists of heating the reaction chamber to a temperature of 94–96 °C (201–205 °F), or 98 °C (208 °F) if extremely thermostable polymerases are used, which is then held for 1–10 minutes.

- Denaturation: This step is the first regular cycling event and consists of heating the reaction chamber to 94–98 °C (201–208 °F) for 20–30 seconds. This causes DNA melting, or denaturation, of the double-stranded DNA template by breaking the hydrogen bonds between complementary bases, yielding two single-stranded DNA molecules.

- Annealing: In the next step, the reaction temperature is lowered to 50–65 °C (122–149 °F) for 20–40 seconds, allowing annealing of the primers to each of the single-stranded DNA templates. Two different primers are typically included in the reaction mixture: one for each of the two single-stranded complements containing the target region. The primers are single-stranded sequences themselves, but are much shorter than the length of the target region, complementing only very short sequences at the 3' end of each strand.

  It is critical to determine a proper temperature for the annealing step because efficiency and specificity are strongly affected by the annealing temperature. This temperature must be low enough to allow for hybridization of the primer to the strand, but high enough for the hybridization to be specific, i.e., the primer should bind *only* to a perfectly complementary part of the strand, and nowhere else. If the temperature is too low, the primer may bind imperfectly. If it is too high, the primer may not bind at all. A typical annealing temperature is about 3–5 °C below the T_m of the primers used. Stable hydrogen bonds between complementary bases are formed only when the primer sequence very closely matches the template sequence. During this step, the polymerase binds to the primer-template hybrid and begins DNA formation.

- Extension/elongation: The temperature at this step depends on the DNA polymerase used; the optimum activity temperature for the thermostable DNA polymerase of Taq (Thermus aquaticus) polymerase is approximately 75–80 °C (167–176 °F), though a temperature of 72 °C (162 °F) is commonly used with this enzyme. In this step, the DNA polymerase synthesizes a new DNA strand complementary to the DNA template strand by adding free dNTPs from the reaction mixture that are complementary to the template in the 5'-to-3' direction, condensing the 5'-phosphate group of the dNTPs with the 3'-hydroxy group at the end of the nascent (elongating) DNA strand. The precise time required for elongation depends both on the DNA polymerase used and on the length of the DNA target region to amplify. As a rule of thumb, at their optimal temperature, most DNA polymerases polymerize a thousand bases per minute. Under optimal conditions (i.e., if there are no limitations due to limiting substrates or reagents), at each extension/elongation step, the number of DNA target sequences is doubled. With each successive cycle, the original template strands plus all newly generated strands become template strands for the next round of elongation, leading to exponential (geometric) amplification of the specific DNA target region.

The processes of denaturation, annealing and elongation constitute a single cycle. Multiple cycles are required to amplify the DNA target to millions of copies. The formula used to calculate the number of DNA copies formed after a given number of cycles is 2^n, where n is the number of cycles. Thus, a reaction set for 30 cycles results in 2^{30}, or 1073741824, copies of the original double-stranded DNA target region.

- Final elongation: This single step is optional, but is performed at a temperature of 70–74 °C (158–165 °F) (the temperature range required for optimal activity of most polymerases used in PCR) for 5–15 minutes after the last PCR cycle to ensure that any remaining single-stranded DNA is fully elongated.

- Final hold: The final step cools the reaction chamber to 4–15 °C (39–59 °F) for an indefinite time, and may be employed for short-term storage of the PCR products.

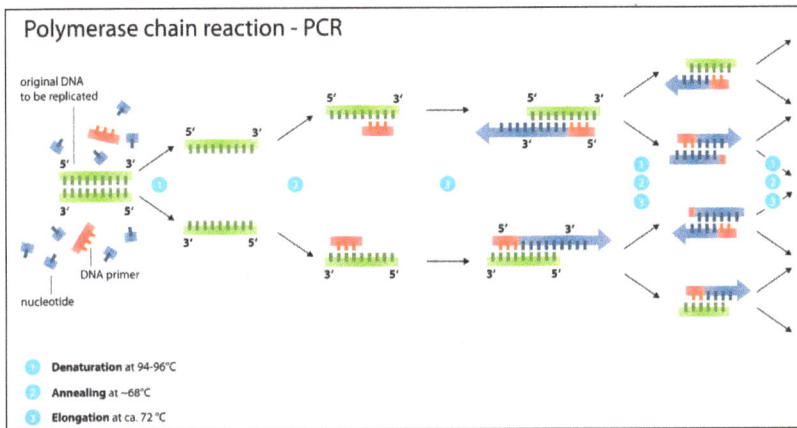

To check whether the PCR successfully generated the anticipated DNA target region (also sometimes referred to as the amplimer or amplicon), agarose gel electrophoresis may be employed for size separation of the PCR products. The size(s) of PCR products is determined by comparison with a DNA ladder, a molecular weight marker which contains DNA fragments of known size run on the gel alongside the PCR products.

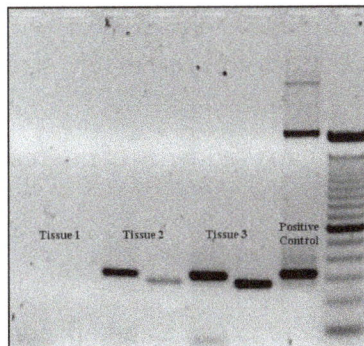

Ethidium bromide-stained PCR products after gel electrophoresis. Two sets of primers were used to amplify a target sequence from three different tissue samples. No amplification is present in sample #1; DNA bands in sample #2 and #3 indicate successful amplification of the target sequence. The gel also shows a positive control, and a DNA ladder containing DNA fragments of defined length for sizing the bands in the experimental PCRs.

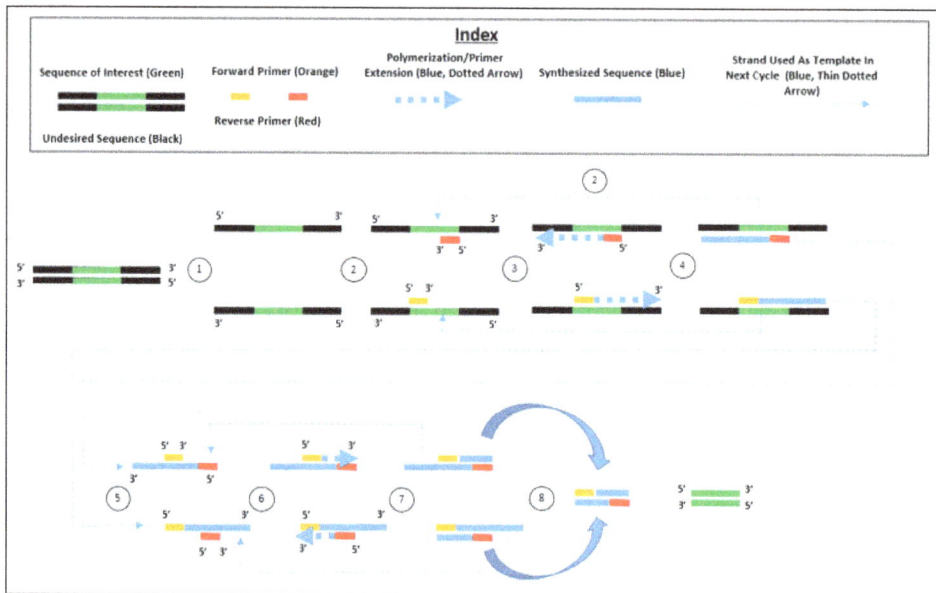

- The DNA double helix is melted apart T> 90 °C and its strands separate.

- The temperature is decreased to slightly below the Tm of both the primers being used. Both primers bind to the available strands. These primers are supplied in excess to insure that the strands do not only come back and reanneal to one another.

- Polymerization (extension) occurs via DNA Polymerase in the 5' to 3' direction on each strand.

- Incorporated additional nucleotides give rise to new strands that extend past the sequence of interest.

- The previously polymerized strands act as template for the other primer (if forward primer bound first, reverse primer now binds and vice versa).

- Polymerization occurs via DNA Polymerase in the 5' to 3' direction on each strands, this time ending at the end of the sequence of interest.

- Incorporated additional nucleotides give rise to new strands that only encode the sequence of interest.

- The synthesized strand encoding the sequence of interest anneal to one another to form the end product.

Stages

As with other chemical reactions, the reaction rate and efficiency of PCR are affected by limiting factors. Thus, the entire PCR process can further be divided into three stages based on reaction progress:

- Exponential amplification: At every cycle, the amount of product is doubled (assuming 100% reaction efficiency). After 30 cycles, a single copy of DNA can be increased up to 1,000,000,000 (one billion) copies. In a sense, then, the replication of a discrete strand of DNA is being manipulated in a tube under controlled conditions. The reaction is very sensitive: only minute quantities of DNA must be present.

- Leveling off stage: The reaction slows as the DNA polymerase loses activity and as consumption of reagents, such as dNTPs and primers, causes them to become more limited.

- Plateau: No more product accumulates due to exhaustion of reagents and enzyme.

Optimization

In practice, PCR can fail for various reasons, in part due to its sensitivity to contamination causing amplification of spurious DNA products. Because of this, a number of techniques and procedures have been developed for optimizing PCR conditions. Contamination with extraneous DNA is addressed with lab protocols and procedures that separate pre-PCR mixtures from potential DNA contaminants. This usually involves spatial separation of PCR-setup areas from areas for analysis or purification of PCR products, use of disposable plasticware, and thoroughly cleaning the work surface between reaction setups. Primer-design techniques are important in improving PCR product yield and in avoiding the formation of spurious products, and the usage of alternate buffer components or polymerase enzymes can help with amplification of long or otherwise problematic regions of DNA. Addition of reagents, such as formamide, in buffer systems may increase the specificity and yield of PCR. Computer simulations of theoretical PCR results (Electronic PCR) may be performed to assist in primer design.

Applications

Selective DNA Isolation

PCR allows isolation of DNA fragments from genomic DNA by selective amplification of a specific region of DNA. This use of PCR augments many ways, such as generating hybridization probes for Southern or northern hybridization and DNA cloning, which require larger amounts of DNA, representing a specific DNA region. PCR supplies these techniques with high amounts of pure DNA, enabling analysis of DNA samples even from very small amounts of starting material.

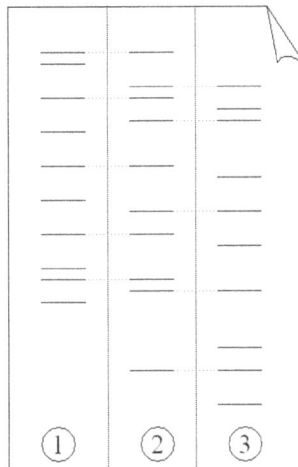

Electrophoresis of PCR-amplified DNA fragments. (1) Father. (2) Child. (3) Mother. The child has inherited some, but not all of the fingerprint of each of its parents, giving it a new, unique fingerprint.

Other applications of PCR include DNA sequencing to determine unknown PCR-amplified sequences in which one of the amplification primers may be used in Sanger sequencing, isolation of a DNA sequence to expedite recombinant DNA technologies involving the insertion of a DNA sequence into a plasmid, phage, or cosmid (depending on size) or the genetic material of another organism. Bacterial colonies *(such as E. coli)* can be rapidly screened by PCR for correct DNA vector constructs. PCR may also be used for genetic fingerprinting; a forensic technique used to identify a person or organism by comparing experimental DNAs through different PCR-based methods.

Some PCR 'fingerprints' methods have high discriminative power and can be used to identify genetic relationships between individuals, such as parent-child or between siblings, and are used in paternity testing (Figure). This technique may also be used to determine evolutionary relationships among organisms when certain molecular clocks are used (i.e., the 16S rRNA and recA genes of microorganisms).

Amplification and Quantification of DNA

Because PCR amplifies the regions of DNA that it targets, PCR can be used to analyze extremely small amounts of sample. This is often critical for forensic analysis, when only a trace amount of DNA is available as evidence. PCR may also be used in the analysis of ancient DNA that is tens of thousands of years old. These PCR-based techniques have been successfully used on animals, such as a forty-thousand-year-old mammoth, and also on human DNA, in applications ranging from the analysis of Egyptian mummies to the identification of a Russian tsar and the body of English king Richard III.

Quantitative PCR or Real Time PCR (qPCR, not to be confused with RT-PCR) methods allow the estimation of the amount of a given sequence present in a sample—a technique often applied to quantitatively determine levels of gene expression. Quantitative PCR is an established tool for DNA quantification that measures the accumulation of DNA product after each round of PCR amplification.

qPCR allows the quantification and detection of a specific DNA sequence in real time since it measures concentration while the synthesis process is taking place. There are two methods for simultaneous detection and quantification. The first method consists of using fluorescent dyes that are retained nonspecifically in between the double strands. The second method involves probes that code for specific sequences and are fluorescently labeled. Detection of DNA using these methods can only be seen after the hybridization of probes with its complementary DNA takes place. An interesting technique combination is real-time PCR and reverse transcription. This sophisticated technique, called RT-qPCR, allows for the quantification of a small quantity of RNA. Through this combined technique, mRNA is converted to cDNA, which is further quantified using qPCR. This technique lowers the possibility of error at the end point of PCR, increasing chances for detection of genes associated with genetic diseases such as cancer. Laboratories use RT-qPCR for the purpose of sensitively measuring gene regulation.

Medical and Diagnostic Applications

Prospective parents can be tested for being genetic carriers, or their children might be tested for actually being affected by a disease. DNA samples for prenatal testing can be obtained by amniocentesis, chorionic villus sampling, or even by the analysis of rare fetal cells circulating in the mother's bloodstream. PCR analysis is also essential to preimplantation genetic diagnosis, where individual cells of a developing embryo are tested for mutations.

- PCR can also be used as part of a sensitive test for *tissue typing*, vital to organ transplantation. As of 2008, there is even a proposal to replace the traditional antibody-based tests for blood type with PCR-based tests.

- Many forms of cancer involve alterations to *oncogenes*. By using PCR-based tests to study these mutations, therapy regimens can sometimes be individually customized to a patient. PCR permits early diagnosis of malignant diseases such as leukemia and lymphomas, which is currently the highest-developed in cancer research and is already being used routinely. PCR assays can be performed directly on genomic DNA samples to detect translocation-specific malignant cells at a sensitivity that is at least 10,000 fold higher than that of other methods. PCR is very useful in the medical field since it allows for the isolation and amplification of tumor suppressors. Quantitative PCR for example, can be used to quantify and analyze single cells, as well as recognize DNA, mRNA and protein confirmations and combinations.

Infectious Disease Applications

PCR allows for rapid and highly specific diagnosis of infectious diseases, including those caused by bacteria or viruses. PCR also permits identification of non-cultivatable or slow-growing microorganisms such as mycobacteria, anaerobic bacteria, or viruses from tissue culture assays and animal models. The basis for PCR diagnostic applications in microbiology is the detection of infectious agents and the discrimination of non-pathogenic from pathogenic strains by virtue of specific genes.

Characterization and detection of infectious disease organisms have been revolutionized by PCR in the following ways:

- The human immunodeficiency virus (or HIV), is a difficult target to find and eradicate. The earliest tests for infection relied on the presence of antibodies to the virus circulating in the bloodstream. However, antibodies don't appear until many weeks after infection, maternal antibodies mask the infection of a newborn, and therapeutic agents to fight the infection don't affect the antibodies. PCR tests have been developed that can detect as little as one viral genome among the DNA of over 50,000 host cells. Infections can be detected earlier, donated blood can be screened directly for the virus, newborns can be immediately tested for infection, and the effects of antiviral treatments can be quantified.

- Some disease organisms, such as that for *tuberculosis*, are difficult to sample from patients and slow to be grown in the laboratory. PCR-based tests have allowed detection of small numbers of disease organisms (both live or dead), in convenient samples. Detailed genetic analysis can also be used to detect antibiotic resistance, allowing immediate and effective therapy. The effects of therapy can also be immediately evaluated.

- The spread of a *disease organism* through populations of domestic or wild animals can be monitored by PCR testing. In many cases, the appearance of new virulent sub-types can be detected and monitored. The sub-types of an organism that were responsible for earlier epidemics can also be determined by PCR analysis.

- Viral DNA can be detected by PCR. The primers used must be specific to the targeted sequences in the DNA of a virus, and PCR can be used for diagnostic analyses or DNA sequencing of the viral genome. The high sensitivity of PCR permits virus detection soon after infection and even before the onset of disease. Such early detection may give physicians a significant lead time in treatment. The amount of virus ("viral load") in a patient can also be quantified by PCR-based DNA quantitation techniques.

- Diseases such as pertussis (or whooping cough) are cause by the bacteria Bordetella pertussis. This bacteria is marked by a serious acute respiratory infection that affects various animals and humans and has led to the deaths of many young children. The pertussis toxin is a protein exotoxin that binds to cell receptors by two dimers and reacts with different cell types such as T lymphocytes which plays a role in cell immunity. PCR is an important testing tool that can detect the sequences that are within the pertussis toxin gene. This is because PCR has a high sensitivity for the toxin and has demonstrated a rapid turnaround time. PCR is very efficient for diagnosing pertussis when compared to culture.

Forensic Applications

The development of PCR-based genetic (or DNA) fingerprinting protocols has seen widespread application in forensics:

- In its most discriminating form, *genetic fingerprinting* can uniquely discriminate any one person from the entire population of the world. Minute samples of DNA can be isolated from a crime scene, and compared to that from suspects, or from a DNA database of earlier evidence or convicts. Simpler versions of these tests are often used to rapidly rule out suspects during a criminal investigation. Evidence from decades-old crimes can be tested, confirming or exonerating the people originally convicted.

- Forensic DNA typing has been an effective way of identifying or exonerating criminal suspects due to analysis of evidence discovered at a crime scene. The human genome has many repetitive regions that can be found within gene sequences or in non-coding regions of the genome. Specifically, up to 40% of human DNA is repetitive. There are two distinct categories for these repetitive, non-coding regions in the genome. The first category is called variable number tandem repeats (VNTR), which are 10–100 base pairs long and the second category is called short tandem repeats (STR) and these consist of repeated 2–10 base pair sections. PCR is used to amplify several well-known VNTRs and STRs using primers that flank each of the repetitive regions. The sizes of the fragments obtained from any individual for each of the STRs will indicate which alleles are present. By analyzing several STRs for an individual, a set of alleles for each person will be found that statistically is likely to be unique. Researchers have identified the complete sequence of the human genome. This sequence can be easily accessed through the NCBI website and is used in many real-life applications. For example, the FBI has compiled a set of DNA marker sites used for identification, and these are called the Combined DNA Index System (CODIS) DNA database. Using this database enables statistical analysis to be used to determine the probability that a DNA sample will match. PCR is a very powerful and significant analytical tool to use for forensic DNA typing because researchers only need a very small amount of the target DNA to be used for analysis. For example, a single human hair with attached hair follicle has enough DNA to conduct the analysis. Similarly, a few sperm, skin samples from under the fingernails, or a small amount of blood can provide enough DNA for conclusive analysis.

- Less discriminating forms of DNA fingerprinting can help in *DNA paternity testing*, where an individual is matched with their close relatives. DNA from unidentified human remains can be tested, and compared with that from possible parents, siblings, or children. Similar testing can be used to confirm the biological parents of an adopted (or kidnapped) child. The actual biological father of a newborn can also be confirmed (or ruled out).

- The PCR AMGX/AMGY design has been shown to not only facilitating in amplifying DNA sequences from a very minuscule amount of genome. However it can also be used for real time sex determination from forensic bone samples. This provides us with a powerful and effective way to determine the sex of not only ancient specimens but also current suspects in crimes.

Research Applications

PCR has been applied to many areas of research in molecular genetics:

- PCR allows rapid production of short pieces of DNA, even when not more than the sequence of the two primers is known. This ability of PCR augments many methods, such as generating *hybridization probes* for Southern or northern blot hybridization. PCR supplies these techniques with large amounts of pure DNA, sometimes as a single strand, enabling analysis even from very small amounts of starting material.

- The task of *DNA sequencing* can also be assisted by PCR. Known segments of DNA can easily be produced from a patient with a genetic disease mutation. Modifications to the amplification technique can extract segments from a completely unknown genome, or can generate just a single strand of an area of interest.

- PCR has numerous applications to the more traditional process of *DNA cloning*. It can extract segments for insertion into a vector from a larger genome, which may be only available in small quantities. Using a single set of 'vector primers', it can also analyze or extract fragments that have already been inserted into vectors. Some alterations to the PCR protocol can *generate mutations* (general or site-directed) of an inserted fragment.

- *Sequence-tagged sites* is a process where PCR is used as an indicator that a particular segment of a genome is present in a particular clone. The Human Genome Project found this application vital to mapping the cosmid clones they were sequencing, and to coordinating the results from different laboratories.

- An exciting application of PCR is the phylogenic analysis of DNA from *ancient sources*, such as that found in the recovered bones of Neanderthals, from frozen tissues of mammoths, or from the brain of Egyptian mummies. Have been amplified and sequenced. In some cases the highly degraded DNA from these sources might be reassembled during the early stages of amplification.

- A common application of PCR is the study of patterns of *gene expression*. Tissues (or even individual cells) can be analyzed at different stages to see which genes have become active, or which have been switched off. This application can also use quantitative PCR to quantitate the actual levels of expression.

- The ability of PCR to simultaneously amplify several loci from individual sperm has greatly enhanced the more traditional task of *genetic mapping* by studying chromosomal crossovers after meiosis. Rare crossover events between very close loci have been directly observed by analyzing thousands of individual sperms. Similarly, unusual deletions, insertions, translocations, or inversions can be analyzed, all without having to wait (or pay) for the long and laborious processes of fertilization, embryogenesis, etc.

- Site-directed mutagenesis: PCR can be used to create mutant genes with mutations chosen by scientists at will. These mutations can be chosen in order to understand how proteins accomplish their functions, and to change or improve protein function.

Advantages

PCR has a number of advantages. It is fairly simple to understand and to use, and produces results rapidly. The technique is highly sensitive with the potential to produce millions to billions of copies of a specific product for sequencing, cloning, and analysis. qRT-PCR shares the same advantages as the PCR, with an added advantage of quantification of the synthesized product. Therefore, it has its uses to analyze alterations of gene expression levels in tumors, microbes, or other disease states.

PCR is a very powerful and practical research tool. The sequencing of unknown etiologies of many diseases are being figured out by the PCR. The technique can help identify the sequence of previously unknown viruses related to those already known and thus give us a better understanding of the disease itself. If the procedure can be further simplified and sensitive non radiometric detection systems can be developed, the PCR will assume a prominent place in the clinical laboratory for years to come.

Limitations

One major limitation of PCR is that prior information about the target sequence is necessary in order to generate the primers that will allow its selective amplification. This means that, typically, PCR users must know the precise sequence upstream of the target region on each of the two single-stranded templates in order to ensure that the DNA polymerase properly binds to the primer-template hybrids and subsequently generates the entire target region during DNA synthesis.

Like all enzymes, DNA polymerases are also prone to error, which in turn causes mutations in the PCR fragments that are generated.

Another limitation of PCR is that even the smallest amount of contaminating DNA can be amplified, resulting in misleading or ambiguous results. To minimize the chance of contamination, investigators should reserve separate rooms for reagent preparation, the PCR, and analysis of product. Reagents should be dispensed into single-use aliquots. Pipetters with disposable plungers and extra-long pipette tips should be routinely used.

Variations

- Allele-specific PCR: a diagnostic or cloning technique based on single-nucleotide variations (SNVs not to be confused with SNPs) (single-base differences in a patient). It requires prior knowledge of a DNA sequence, including differences between alleles, and uses primers whose 3' ends encompass the SNV (base pair buffer around SNV usually incorporated). PCR amplification under stringent conditions is much less efficient in the presence of a mismatch between template and primer, so successful amplification with an SNP-specific primer signals presence of the specific SNP in a sequence.

- Assembly PCR or Polymerase Cycling Assembly (PCA): artificial synthesis of long DNA sequences by performing PCR on a pool of long oligonucleotides with short overlapping segments. The oligonucleotides alternate between sense and antisense directions, and the overlapping segments determine the order of the PCR fragments, thereby selectively producing the final long DNA product.

- Asymmetric PCR: preferentially amplifies one DNA strand in a double-stranded DNA template. It is used in sequencing and hybridization probing where amplification of only one of the two complementary strands is required. PCR is carried out as usual, but with a great excess of the primer for the strand targeted for amplification. Because of the slow (arithmetic) amplification later in the reaction after the limiting primer has been used up, extra cycles of PCR are required. A recent modification on this process, known as Linear-After-The-Exponential-PCR (LATE-PCR), uses a limiting primer with a higher melting temperature (Tm) than the excess primer to maintain reaction efficiency as the limiting primer concentration decreases mid-reaction.

- Convective PCR: a pseudo-isothermal way of performing PCR. Instead of repeatedly heating and cooling the PCR mixture, the solution is subjected to a thermal gradient. The resulting thermal instability driven convective flow automatically shuffles the PCR reagents from the hot and cold regions repeatedly enabling PCR. Parameters such as thermal boundary conditions and geometry of the PCR enclosure can be optimized to yield robust and rapid PCR by harnessing the emergence of chaotic flow fields. Such convective flow PCR setup significantly reduces device power requirement and operation time.

- Dial-out PCR: a highly parallel method for retrieving accurate DNA molecules for gene synthesis. A complex library of DNA molecules is modified with unique flanking tags before massively parallel sequencing. Tag-directed primers then enable the retrieval of molecules with desired sequences by PCR.

- Digital PCR (dPCR): used to measure the quantity of a target DNA sequence in a DNA sample. The DNA sample is highly diluted so that after running many PCRs in parallel, some of them do not receive a single molecule of the target DNA. The target DNA concentration is calculated using the proportion of negative outcomes. Hence the name 'digital PCR'.

- Helicase-dependent amplification: similar to traditional PCR, but uses a constant temperature rather than cycling through denaturation and annealing/extension cycles. DNA helicase, an enzyme that unwinds DNA, is used in place of thermal denaturation.

- Hot start PCR: a technique that reduces non-specific amplification during the initial set up stages of the PCR. It may be performed manually by heating the reaction components to the denaturation temperature (e.g., 95 °C) before adding the polymerase. Specialized enzyme systems have been developed that inhibit the polymerase's activity at ambient temperature, either by the binding of an antibody or by the presence of covalently bound inhibitors that dissociate only after a high-temperature activation step. Hot-start/cold-finish PCR is achieved with new hybrid polymerases that are inactive at ambient temperature and are instantly activated at elongation temperature.

- In silico PCR (digital PCR, virtual PCR, electronic PCR, e-PCR) refers to computational tools used to calculate theoretical polymerase chain reaction results using a given set of primers (probes) to amplify DNA sequences from a sequenced genome or transcriptome. In silico PCR was proposed as an educational tool for molecular biology.

- Intersequence-specific PCR (ISSR): a PCR method for DNA fingerprinting that amplifies regions between simple sequence repeats to produce a unique fingerprint of amplified fragment lengths.

- Inverse PCR: is commonly used to identify the flanking sequences around genomic inserts. It involves a series of DNA digestions and self ligation, resulting in known sequences at either end of the unknown sequence.

- Ligation-mediated PCR: uses small DNA linkers ligated to the DNA of interest and multiple primers annealing to the DNA linkers; it has been used for DNA sequencing, genome walking, and DNA footprinting.

- Methylation-specific PCR (MSP): developed by Stephen Baylin and James G. Herman at the Johns Hopkins School of Medicine, and is used to detect methylation of CpG islands in genomic DNA. DNA is first treated with sodium bisulfite, which converts unmethylated cytosine bases to uracil, which is recognized by PCR primers as thymine. Two PCRs are then carried out on the modified DNA, using primer sets identical except at any CpG islands within the primer sequences. At these points, one primer set recognizes DNA with cytosines to amplify methylated DNA, and one set recognizes DNA with uracil or thymine to amplify unmethylated DNA. MSP using qPCR can also be performed to obtain quantitative rather than qualitative information about methylation.

- Miniprimer PCR: uses a thermostable polymerase (S-Tbr) that can extend from short primers ("smalligos") as short as 9 or 10 nucleotides. This method permits PCR targeting to smaller primer binding regions, and is used to amplify conserved DNA sequences, such as the 16S (or eukaryotic 18S) rRNA gene.

- Multiplex ligation-dependent probe amplification (MLPA): permits amplifying multiple targets with a single primer pair, thus avoiding the resolution limitations of multiplex PCR.

- Multiplex-PCR: consists of multiple primer sets within a single PCR mixture to produce amplicons of varying sizes that are specific to different DNA sequences. By targeting multiple genes at once, additional information may be gained from a single test-run that otherwise would require several times the reagents and more time to perform. Annealing temperatures for each of the primer sets must be optimized to work correctly within a single reaction, and amplicon sizes. That is, their base pair length should be different enough to form distinct bands when visualized by gel electrophoresis.

- Nanoparticle-Assisted PCR (nanoPCR): some nanoparticles (NPs) can enhance the efficiency of PCR (thus being called nanoPCR), and some can even outperform the original

PCR enhancers. It was reported that quantum dots (QDs) can improve PCR specificity and efficiency. Single-walled carbon nanotubes (SWCNTs) and multi-walled carbon nanotubes (MWCNTs) are efficient in enhancing the amplification of long PCR. Carbon nanopowder (CNP) can improve the efficiency of repeated PCR and long PCR, while zinc oxide, titanium dioxide and Ag NPs were found to increase the PCR yield. Previous data indicated that non-metallic NPs retained acceptable amplification fidelity. Given that many NPs are capable of enhancing PCR efficiency, it is clear that there is likely to be great potential for nanoPCR technology improvements and product development.

- Nested PCR: increases the specificity of DNA amplification, by reducing background due to non-specific amplification of DNA. Two sets of primers are used in two successive PCRs. In the first reaction, one pair of primers is used to generate DNA products, which besides the intended target, may still consist of non-specifically amplified DNA fragments. The product(s) are then used in a second PCR with a set of primers whose binding sites are completely or partially different from and located 3' of each of the primers used in the first reaction. Nested PCR is often more successful in specifically amplifying long DNA fragments than conventional PCR, but it requires more detailed knowledge of the target sequences.

- Overlap-extension PCR or Splicing by overlap extension (SOEing) : a genetic engineering technique that is used to splice together two or more DNA fragments that contain complementary sequences. It is used to join DNA pieces containing genes, regulatory sequences, or mutations; the technique enables creation of specific and long DNA constructs. It can also introduce deletions, insertions or point mutations into a DNA sequence.

- PAN-AC: uses isothermal conditions for amplification, and may be used in living cells.

- quantitative PCR (qPCR): used to measure the quantity of a target sequence (commonly in real-time). It quantitatively measures starting amounts of DNA, cDNA, or RNA. quantitative PCR is commonly used to determine whether a DNA sequence is present in a sample and the number of its copies in the sample. Quantitative PCR has a very high degree of precision. Quantitative PCR methods use fluorescent dyes, such as Sybr Green, EvaGreen or fluorophore-containing DNA probes, such as TaqMan, to measure the amount of amplified product in real time. It is also sometimes abbreviated to RT-PCR (real-time PCR) but this abbreviation should be used only for reverse transcription PCR. qPCR is the appropriate contractions for quantitative PCR (real-time PCR).

- Reverse Transcription PCR (RT-PCR): for amplifying DNA from RNA. Reverse transcriptase reverse transcribes RNA into cDNA, which is then amplified by PCR. RT-PCR is widely used in expression profiling, to determine the expression of a gene or to identify the sequence of an RNA transcript, including transcription start and termination sites. If the genomic DNA sequence of a gene is known, RT-PCR can be used to map the location of exons and introns in the gene. The 5' end of a gene (corresponding to the transcription start site) is typically identified by RACE-PCR (Rapid Amplification of cDNA Ends).

- RNase H-dependent PCR (rhPCR): a modification of PCR that utilizes primers with a 3' extension block that can be removed by a thermostable RNase HII enzyme. This system reduces primer-dimers and allows for multiplexed reactions to be performed with higher numbers of primers.

- Single Specific Primer-PCR (SSP-PCR): allows the amplification of double-stranded DNA even when the sequence information is available at one end only. This method permits amplification of genes for which only a partial sequence information is available, and allows unidirectional genome walking from known into unknown regions of the chromosome.

- Solid Phase PCR: encompasses multiple meanings, including Polony Amplification (where PCR colonies are derived in a gel matrix, for example), Bridge PCR (primers are covalently linked to a solid-support surface), conventional Solid Phase PCR (where Asymmetric PCR is applied in the presence of solid support bearing primer with sequence matching one of the aqueous primers) and Enhanced Solid Phase PCR (where conventional Solid Phase PCR can be improved by employing high Tm and nested solid support primer with optional application of a thermal 'step' to favour solid support priming).

- Suicide PCR: typically used in paleogenetics or other studies where avoiding false positives and ensuring the specificity of the amplified fragment is the highest priority. It was originally described in a study to verify the presence of the microbe Yersinia pestis in dental samples obtained from 14th Century graves of people supposedly killed by plague during the medieval Black Death epidemic. The method prescribes the use of any primer combination only once in a PCR (hence the term "suicide"), which should never have been used in any positive control PCR reaction, and the primers should always target a genomic region never amplified before in the lab using this or any other set of primers. This ensures that no contaminating DNA from previous PCR reactions is present in the lab, which could otherwise generate false positives.

- Thermal asymmetric interlaced PCR (TAIL-PCR): for isolation of an unknown sequence flanking a known sequence. Within the known sequence, TAIL-PCR uses a nested pair of primers with differing annealing temperatures; a degenerate primer is used to amplify in the other direction from the unknown sequence.

- Touchdown PCR (Step-down PCR): a variant of PCR that aims to reduce nonspecific background by gradually lowering the annealing temperature as PCR cycling progresses. The annealing temperature at the initial cycles is usually a few degrees (3–5 °C) above the Tm of the primers used, while at the later cycles, it is a few degrees (3–5 °C) below the primer Tm. The higher temperatures give greater specificity for primer binding, and the lower temperatures permit more efficient amplification from the specific products formed during the initial cycles.

- Universal Fast Walking: for genome walking and genetic fingerprinting using a more specific 'two-sided' PCR than conventional 'one-sided' approaches (using only one gene-specific primer and one general primer—which can lead to artefactual 'noise') by virtue of a mechanism involving lariat structure formation. Streamlined derivatives of UFW are LaNe RAGE (lariat-dependent nested PCR for rapid amplification of genomic DNA ends), 5'RACE LaNe and 3'RACE LaNe.

Southern Blot

A Southern blot is a method used in molecular biology for detection of a specific DNA sequence in DNA samples. Southern blotting combines transfer of electrophoresis-separated DNA fragments to a filter membrane and subsequent fragment detection by probe hybridization.

The method is named after the British biologist Edwin Southern, who first published it in 1975. Other blotting methods (i.e., western blot, northern blot, eastern blot, southwestern blot) that employ similar principles, but using RNA or protein, have later been named in reference to Edwin Southern's name. As the label is eponymous, Southern is capitalised, as is conventional of proper nouns. The names for other blotting methods may follow this convention, by analogy.

Method

1. Restriction endonucleases are used to cut high-molecular-weight DNA strands into smaller fragments.

2. The DNA fragments are then electrophoresed on an agarose gel to separate them by size.

3. If some of the DNA fragments are larger than 15 kb, then prior to blotting, the gel may be treated with an acid, such as dilute HCl. This depurinates the DNA fragments, breaking the DNA into smaller pieces, thereby allowing more efficient transfer from the gel to membrane.

4. If alkaline transfer methods are used, the DNA gel is placed into an alkaline solution (typically containing sodium hydroxide) to denature the double-stranded DNA. The denaturation in an alkaline environment may improve binding of the negatively charged thymine residues of DNA to a positively charged amino groups of membrane, separating it into single DNA strands for later hybridization to the probe, and destroys any residual RNA that may still be present in the DNA. The choice of alkaline over neutral transfer methods, however, is often empirical and may result in equivalent results.

5. A sheet of nitrocellulose (or, alternatively, nylon) membrane is placed on top of (or below, depending on the direction of the transfer) the gel. Pressure is applied evenly to the gel (either using suction, or by placing a stack of paper towels and a weight on top of the membrane and gel), to ensure good and even contact between gel and membrane. If transferring by suction, 20X SSC buffer is used to ensure a seal and prevent drying of the gel. Buffer transfer by capillary action from a region of high water potential to a region of low water potential (usually filter paper and paper tissues) is then used to move the DNA from the gel onto the membrane; ion exchange interactions bind the DNA to the membrane due to the negative charge of the DNA and positive charge of the membrane.

6. The membrane is then baked in a vacuum or regular oven at 80 °C for 2 hours (standard conditions; nitrocellulose or nylon membrane) or exposed to ultraviolet radiation (nylon membrane) to permanently attach the transferred DNA to the membrane.

7. The membrane is then exposed to a hybridization probe—a single DNA fragment with a specific sequence whose presence in the target DNA is to be determined. The probe DNA is labelled so that it can be detected, usually by incorporating radioactivity or tagging the

molecule with a fluorescent or chromogenic dye. In some cases, the hybridization probe may be made from RNA, rather than DNA. To ensure the specificity of the binding of the probe to the sample DNA, most common hybridization methods use salmon or herring sperm DNA for blocking of the membrane surface and target DNA, deionized formamide, and detergents such as SDS to reduce non-specific binding of the probe.

8. After hybridization, excess probe is washed from the membrane (typically using SSC buffer), and the pattern of hybridization is visualized on X-ray film by autoradiography in the case of a radioactive or fluorescent probe, or by development of colour on the membrane if a chromogenic detection method is used.

Result

Hybridization of the probe to a specific DNA fragment on the filter membrane indicates that this fragment contains DNA sequence that is complementary to the probe. The transfer step of the DNA from the electrophoresis gel to a membrane permits easy binding of the labeled hybridization probe to the size-fractionated DNA. It also allows for the fixation of the target-probe hybrids, required for analysis by autoradiography or other detection methods. Southern blots performed with restriction enzyme-digested genomic DNA may be used to determine the number of sequences (e.g., gene copies) in a genome. A probe that hybridizes only to a single DNA segment that has not been cut by the restriction enzyme will produce a single band on a Southern blot, whereas multiple bands will likely be observed when the probe hybridizes to several highly similar sequences (e.g., those that may be the result of sequence duplication). Modification of the hybridization conditions (for example, increasing the hybridization temperature or decreasing salt concentration) may be used to increase specificity and decrease hybridization of the probe to sequences that are less than 100% similar.

Applications

Southern blotting transfer may be used for homology-based cloning on the basis of amino acid sequence of the protein product of the target gene. Oligonucleotides are designed so that they are similar to the target sequence. The oligonucleotides are chemically synthesized, radiolabeled, and used to screen a DNA library, or other collections of cloned DNA fragments. Sequences that hybridize with the hybridization probe are further analysed, for example, to obtain the full length sequence of the targeted gene.

Southern blotting can also be used to identify methylated sites in particular genes. Particularly useful are the restriction nucleases *MspI* and *HpaII*, both of which recognize and cleave within the same sequence. However, *HpaII* requires that a C within that site be methylated, whereas *MspI* cleaves only DNA unmethylated at that site. Therefore, any methylated sites within a sequence analyzed with a particular probe will be cleaved by the former, but not the latter, enzyme.

Northern Blot

The northern blot, or RNA blot, is a technique used in molecular biology research to study gene expression by detection of RNA (or isolated mRNA) in a sample.

With northern blotting it is possible to observe cellular control over structure and function by determining the particular gene expression rates during differentiation and morphogenesis, as well as in abnormal or diseased conditions. Northern blotting involves the use of electrophoresis to separate RNA samples by size, and detection with a hybridization probe complementary to part of or the entire target sequence. The term 'northern blot' actually refers specifically to the capillary transfer of RNA from the electrophoresis gel to the blotting membrane. However, the entire process is commonly referred to as northern blotting. The northern blot technique was developed in 1977 by James Alwine, David Kemp, and George Stark at Stanford University, with contributions from Gerhard Heinrich. Northern blotting takes its name from its similarity to the first blotting technique, the Southern blot, named for biologist Edwin Southern. The major difference is that RNA, rather than DNA, is analyzed in the northern blot.

Procedure

A general blotting procedure starts with extraction of total RNA from a homogenized tissue sample or from cells. Eukaryotic mRNA can then be isolated through the use of oligo (dT) cellulose chromatography to isolate only those RNAs with a poly(A) tail. RNA samples are then separated by gel electrophoresis. Since the gels are fragile and the probes are unable to enter the matrix, the RNA samples, now separated by size, are transferred to a nylon membrane through a capillary or vacuum blotting system.

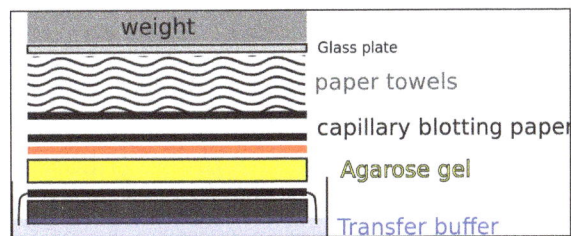

Capillary blotting system setup for the transfer of RNA from an electrophoresis gel to a blotting membrane.

A nylon membrane with a positive charge is the most effective for use in northern blotting since the negatively charged nucleic acids have a high affinity for them. The transfer buffer used for the blotting usually contains formamide because it lowers the annealing temperature of the probe-RNA interaction, thus eliminating the need for high temperatures, which could cause RNA degradation. Once the RNA has been transferred to the membrane, it is immobilized through covalent linkage to the membrane by UV light or heat. After a probe has been labeled, it is hybridized to the RNA on the membrane. Experimental conditions that can affect the efficiency and specificity of hybridization include ionic strength, viscosity, duplex length, mismatched base pairs, and base composition. The membrane is washed to ensure that the probe has bound specifically and to prevent background signals from arising. The hybrid signals are then detected by X-ray film and can be quantified by densitometry. To create controls for comparison in a northern blot, samples not displaying the gene product of interest can be used after determination by microarrays or RT-PCR.

Gels

The RNA samples are most commonly separated on agarose gels containing formaldehyde as a denaturing agent for the RNA to limit secondary structure. The gels can be stained with ethidium bromide (EtBr) and viewed under UV light to observe the quality and quantity of RNA before blotting. Polyacrylamide gel electrophoresis with urea can also be used in RNA separation but it is most commonly used for fragmented RNA or microRNAs. An RNA ladder is often run alongside the samples on an electrophoresis gel to observe the size of fragments obtained but in total RNA samples the ribosomal subunits can act as size markers. Since the large ribosomal subunit is 28S (approximately 5kb) and the small ribosomal subunit is 18S (approximately 2kb) two prominent bands appear on the gel, the larger at close to twice the intensity of the smaller.

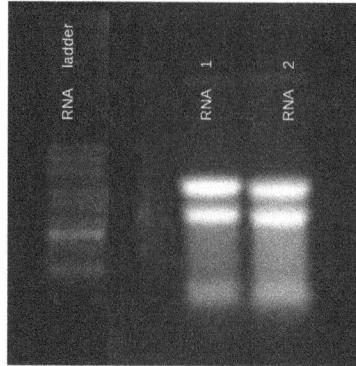

RNA run on a formaldehyde agarose gel to highlight the 28S
(top band) and 18S (lower band) ribosomal subunits.

Probes

Probes for northern blotting are composed of nucleic acids with a complementary sequence to all or part of the RNA of interest, they can be DNA, RNA, or oligonucleotides with a minimum of 25 complementary bases to the target sequence. RNA probes (riboprobes) that are transcribed in vitro are able to withstand more rigorous washing steps preventing some of the background noise. Commonly cDNA is created with labelled primers for the RNA sequence of interest to act as the probe in the northern blot. The probes must be labelled either with radioactive isotopes (^{32}P) or with chemiluminescence in which alkaline phosphatase or horseradish peroxidase (HRP) break down chemiluminescent substrates producing a detectable emission of light. The chemiluminescent labelling can occur in two ways: either the probe is attached to the enzyme, or the probe is labelled with a ligand (e.g. biotin) for which the ligand (e.g., avidin or streptavidin) is attached to the enzyme (e.g. HRP). X-ray film can detect both the radioactive and chemiluminescent signals and many researchers prefer the chemiluminescent signals because they are faster, more sensitive, and reduce the health hazards that go along with radioactive labels. The same membrane can be probed up to five times without a significant loss of the target RNA.

Applications

Northern blotting allows one to observe a particular gene's expression pattern between tissues, organs, developmental stages, environmental stress levels, pathogen infection, and over the course of treatment. The technique has been used to show overexpression of oncogenes and downregulation of tumor-suppressor genes in cancerous cells when compared to 'normal' tissue, as well as the gene expression in the rejection of transplanted organs. If an upregulated gene is observed by an abundance of mRNA on the northern blot the sample can then be sequenced to determine if the gene is known to researchers or if it is a novel finding. The expression patterns obtained under given conditions can provide insight into the function of that gene. Since the RNA is first separated by size, if only one probe type is used variance in the level of each band on the membrane can provide insight into the size of the product, suggesting alternative splice products of the same gene or repetitive sequence motifs. The variance in size of a gene product can also indicate deletions or errors in transcript processing. By altering the probe target used along the known sequence it is possible to determine which region of the RNA is missing.

BlotBase is an online database publishing northern blots. BlotBase has over 700 published northern blots of human and mouse samples, in over 650 genes across more than 25 different tissue types. Northern blots can be searched by a blot ID, paper reference, gene identifier, or by tissue. The results of a search provide the blot ID, species, tissue, gene, expression level, blot image (if available), and links to the publication that the work originated from. This new database provides sharing of information between members of the science community that was not previously seen in northern blotting as it was in sequence analysis, genome determination, protein structure, etc.

Advantages and Disadvantages

Analysis of gene expression can be done by several different methods including RT-PCR, RNase protection assays, microarrays, RNA-Seq, serial analysis of gene expression (SAGE), as well as northern blotting. Microarrays are quite commonly used and are usually consistent with data obtained from northern blots; however, at times northern blotting is able to detect small changes in gene expression that microarrays cannot. The advantage that microarrays have over northern blots is that thousands of genes can be visualized at a time, while northern blotting is usually looking at one or a small number of genes.

A problem in northern blotting is often sample degradation by RNases (both endogenous to the sample and through environmental contamination), which can be avoided by proper sterilization of glassware and the use of RNase inhibitors such as DEPC (diethylpyrocarbonate). The chemicals used in most northern blots can be a risk to the researcher, since formaldehyde, radioactive material, ethidium bromide, DEPC, and UV light are all harmful under certain exposures. Compared to RT-PCR, northern blotting has a low sensitivity, but it also has a high specificity, which is important to reduce false positive results.

The advantages of using northern blotting include the detection of RNA size, the observation of alternate splice products, the use of probes with partial homology, the quality and quantity of RNA can be measured on the gel prior to blotting, and the membranes can be stored and reprobed for years after blotting.

For northern blotting for the detection of acetylcholinesterase mRNA the nonradioactive technique was compared to a radioactive technique and found as sensitive as the radioactive one, but requires no protection against radiation and is less time consuming.

Reverse Northern Blot

Researchers occasionally use a variant of the procedure known as the reverse northern blot. In this procedure, the substrate nucleic acid (that is affixed to the membrane) is a collection of isolated DNA fragments, and the probe is RNA extracted from a tissue and radioactively labelled. The use of DNA microarrays that have come into widespread use in the late 1990s and early 2000s is more akin to the reverse procedure, in that they involve the use of isolated DNA fragments affixed to a substrate, and hybridization with a probe made from cellular RNA. Thus the reverse procedure, though originally uncommon, enabled northern analysis to evolve into gene expression profiling, in which many (possibly all) of the genes in an organism may have their expression monitored.

Interactome

An interactome is the whole set of molecular interactions in a particular cell. The term specifically refers to physical interactions among molecules (such as those among proteins, also known as protein–protein interactions, PPIs; or between small molecules and proteins) but can also describe sets of indirect interactions among genes (genetic interactions). The interactomes based on PPIs should be associated to the proteome of the corresponding species in order to provide a global view ("omic") of all the possible molecular interactions that a protein can present.

The word "interactome" was originally coined in 1999 by a group of French scientists headed by Bernard Jacq. Mathematically, interactomes are generally displayed as graphs. Though interactomes may be described as biological networks, they should not be confused with other networks such as neural networks or food webs.

Molecular Interaction Networks

Molecular interactions can occur between molecules belonging to different biochemical families (proteins, nucleic acids, lipids, carbohydrates, etc.) and also within a given family. Whenever such molecules are connected by physical interactions, they form molecular interaction networks that are generally classified by the nature of the compounds involved. Most commonly, interactome refers to protein–protein interaction (PPI) network (PIN) or subsets thereof. For instance, the Sirt-1 protein interactome and Sirt family second order interactome is the network involving Sirt-1 and its directly interacting proteins where as second order interactome illustrates interactions up to second order of neighbors (Neighbors of neighbors). Another extensively studied type of interactome is the protein–DNA interactome, also called a gene-regulatory network, a network formed by transcription factors, chromatin regulatory proteins, and their target genes. Even metabolic networks can be considered as molecular interaction networks: metabolites, i.e. chemical compounds in a cell, are converted into each other by enzymes, which have to bind their substrates physically.

In fact, all interactome types are interconnected. For instance, protein interactomes contain many enzymes which in turn form biochemical networks. Similarly, gene regulatory networks overlap substantially with protein interaction networks and signaling networks.

Size

It has been suggested that the size of an organism's interactome correlates better than genome size with the biological complexity of the organism. Although protein–protein interaction maps containing several thousand binary interactions are now available for several species, none of them is presently complete and the size of interactomes is still a matter of debate.

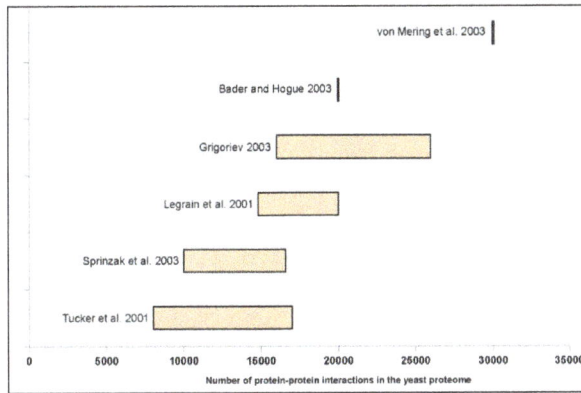

Estimates of the yeast protein interactome.

Yeast

The yeast interactome, i.e. all protein–protein interactions among proteins of *Saccharomyces cerevisiae*, has been estimated to contain between 10,000 and 30,000 interactions. A reasonable estimate may be on the order of 20,000 interactions. Larger estimates often include indirect or predicted interactions, often from affinity purification/mass spectrometry (AP/MS) studies.

Genetic Interaction Networks

Genes interact in the sense that they affect each other's function. For instance, a mutation may be harmless, but when it is combined with another mutation, the combination may turn out to be lethal. Such genes are said to "interact genetically". Genes that are connected in such a way form genetic interaction networks. Some of the goals of these networks are: develop a functional map of a cell's processes, drug target identification, and to predict the function of uncharacterized genes.

In 2010, the most "complete" gene interactome produced to date was compiled from about 5.4 million two-gene comparisons to describe "the interaction profiles for ~75% of all genes in the budding yeast", with ~170,000 gene interactions. The genes were grouped based on similar function so as to build a functional map of the cell's processes. Using this method the study was able to predict known gene functions better than any other genome-scale data set as well as adding functional information for genes that hadn't been previously described. From this model genetic interactions can be observed at multiple scales which will assist in the study of concepts such as gene conservation. Some of the observations made from this study are that there were twice as many negative as positive interactions, negative interactions were more informative than positive interactions, and genes with more connections were more likely to result in lethality when disrupted.

Interactomics

Interactomics is a discipline at the intersection of bioinformatics and biology that deals with studying both the interactions and the consequences of those interactions between and among proteins, and other molecules within a cell. Interactomics thus aims to compare such networks of interactions (i.e., interactomes) between and within species in order to find how the traits of such networks are either preserved or varied.

Interactomics is an example of "top-down" systems biology, which takes an overhead, as well as overall, view of a biosystem or organism. Large sets of genome-wide and proteomic data are collected, and correlations between different molecules are inferred. From the data new hypotheses are formulated about feedbacks between these molecules. These hypotheses can then be tested by new experiments.

Experimental Methods to Map Interactomes

The study of interactomes is called interactomics. The basic unit of a protein network is the protein–protein interaction (PPI). While there are numerous methods to study PPIs, there are relatively few that have been used on a large scale to map whole interactomes.

The yeast two hybrid system (Y2H) is suited to explore the binary interactions among two proteins at a time. Affinity purification and subsequent mass spectrometry is suited to identify a protein complex. Both methods can be used in a high-throughput (HTP) fashion. Yeast two hybrid screens allow false positive interactions between proteins that are never expressed in the same time and place; affinity capture mass spectrometry does not have this drawback, and is the current gold standard. Yeast two-hybrid data better indicates non-specific tendencies towards sticky interactions rather while affinity capture mass spectrometry better indicates functional in vivo protein–protein interactions.

Computational Methods to Study Interactomes

Once an interactome has been created, there are numerous ways to analyze its properties. However, there are two important goals of such analyses. First, scientists try to elucidate the systems properties of interactomes, e.g. the topology of its interactions. Second, studies may focus on individual proteins and their role in the network. Such analyses are mainly carried out using bioinformatics methods and include the following, among many others:

Validation

First, the coverage and quality of an interactome has to be evaluated. Interactomes are never complete, given the limitations of experimental methods. For instance, it has been estimated that typical Y2H screens detect only 25% or so of all interactions in an interactome. The coverage of an interactome can be assessed by comparing it to benchmarks of well-known interactions that have been found and validated by independent assays. Other methods filter out false positives calculating the similarity of known annotations of the proteins involved or define a likelihood of interaction using the subcellular localization of these proteins.

Predicting PPIs

Using experimental data as a starting point, homology transfer is one way to predict interactomes. Here, PPIs from one organism are used to predict interactions among homologous proteins in another organism ("interologs"). However, this approach has certain limitations, primarily because the source data may not be reliable (e.g. contain false positives and false negatives). In addition, proteins and their interactions change during evolution and thus may have been lost or gained. Nevertheless, numerous interactomes have been predicted, e.g. that of Bacillus licheniformis.

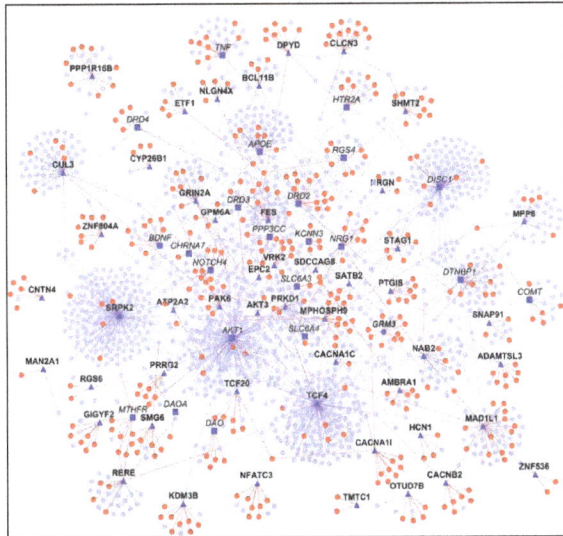

Schziophrenia PPI.

Some algorithms use experimental evidence on structural complexes, the atomic details of binding interfaces and produce detailed atomic models of protein–protein complexes as well as other protein–molecule interactions. Other algorithms use only sequence information, thereby creating unbiased complete networks of interaction with many mistakes.

Some methods use machine learning to distinguish how interacting protein pairs differ from non-interacting protein pairs in terms of pairwise features such as cellular colocalization, gene co-expression, how closely located on a DNA are the genes that encode the two proteins, and so on. Random Forest has been found to be most-effective machine learning method for protein interaction prediction. Such methods have been applied for discovering protein interactions on human interactome, specifically the interactome of Membrane proteins and the interactome of Schizophrenia-associated proteins.

Text Mining of PPIs

Some efforts have been made to extract systematically interaction networks directly from the scientific literature. Such approaches range in terms of complexity from simple co-occurrence statistics of entities that are mentioned together in the same context (e.g. sentence) to sophisticated natural language processing and machine learning methods for detecting interaction relationships.

Protein Function Prediction

Protein interaction networks have been used to predict the function of proteins of unknown functions. This is usually based on the assumption that uncharacterized proteins have similar functions as their interacting proteins (guilt by association). For example, YbeB, a protein of unknown function was found to interact with ribosomal proteins and later shown to be involved in translation. Although such predictions may be based on single interactions, usually several interactions are found. Thus, the whole network of interactions can be used to predict protein functions, given that certain functions are usually enriched among the interactors. The term hypothome has been used to denote an interactome wherein at least one of the genes or proteins is a hypothetical protein.

Perturbations and Disease

The *topology* of an interactome makes certain predictions how a network reacts to the perturbation (e.g. removal) of nodes (proteins) or edges (interactions). Such perturbations can be caused by mutations of genes, and thus their proteins, and a network reaction can manifest as a disease. A network analysis can identify drug targets and biomarkers of diseases.

Network structure and Topology

Interaction networks can be analyzed using the tools of graph theory. Network properties include the degree distribution, clustering coefficients, betweenness centrality, and many others. The distribution of properties among the proteins of an interactome has revealed that the interactome networks often have scale-free topology where functional modules within a network indicate specialized subnetworks. Such modules can be functional, as in a signaling pathway, or structural, as in a protein complex. In fact, it is a formidable task to identify protein complexes in an interactome, given that a network on its own does not directly reveal the presence of a stable complex.

Studied Interactomes

Viral Interactomes

Viral protein interactomes consist of interactions among viral or phage proteins. They were among the first interactome projects as their genomes are small and all proteins can be analyzed with limited resources. Viral interactomes are connected to their host interactomes, forming virus-host interaction networks. Some published virus interactomes include

Bacteriophage

- Escherichia coli bacteriophage lambda.
- Escherichia coli bacteriophage T7.
- Streptococcus pneumoniae bacteriophage Dp-1.
- Streptococcus pneumoniae bacteriophage Cp-1.

The lambda and VZV interactomes are not only relevant for the biology of these viruses but also for technical reasons: they were the first interactomes that were mapped with multiple Y2H vectors, proving an improved strategy to investigate interactomes more completely than previous attempts have shown.

Human (Mammalian) Viruses

- Human varicella zoster virus (VZV).
- Chandipura virus.
- Epstein-Barr virus (EBV).

- Hepatitis C virus (HPC), Human-HCV interactions.

- Hepatitis E virus (HEV).

- Herpes simplex virus 1 (HSV-1).

- Kaposi's sarcoma-associated herpesvirus (KSHV).

- Murine cytomegalovirus (mCMV).

Bacterial Interactomes

Relatively few bacteria have been comprehensively studied for their protein–protein interactions. However, none of these interactomes are complete in the sense that they captured all interactions. In fact, it has been estimated that none of them covers more than 20% or 30% of all interactions, primarily because most of these studies have only employed a single method, all of which discover only a subset of interactions. Among the published bacterial interactomes (including partial ones) are:

Species	Proteins total	Interactions	type
Helicobacter pylori	1,553	~3,004	Y2H
Campylobacter jejuni	1,623	11,687	Y2H
Treponema pallidum	1,040	3,649	Y2H
Escherichia coli	4,288	(5,993)	AP/MS
Escherichia coli	4,288	2,234	Y2H
Mesorhizobium loti	6,752	3,121	Y2H
Mycobacterium tuberculosis	3,959	>8000	B2H
Mycoplasma genitalium	482		AP/MS
Synechocystis sp. PCC6803	3,264	3,236	Y2H
Staphylococcus aureus (MRSA)	2,656	13,219	AP/MS

The *E. coli* and *Mycoplasma* interactomes have been analyzed using large-scale protein complex affinity purification and mass spectrometry (AP/MS), hence it is not easily possible to infer direct interactions. The others have used extensive yeast two-hybrid (Y2H) screens. The *Mycobacterium tuberculosis* interactome has been analyzed using a bacterial two-hybrid screen (B2H).

Note that numerous additional interactomes have been predicted using computational methods.

Eukaryotic Interactomes

There have been several efforts to map eukaryotic interactomes through HTP methods. While no biological interactomes have been fully characterized, over 90% of proteins in *Saccharomyces cerevisiae* have been screened and their interactions characterized, making it the best-characterized interactome. Species whose interactomes have been studied in some detail include

- Schizosaccharomyces pombe.

- Caenorhabditis elegans.

- Drosophila melanogaster.

- Homo sapiens.

Recently, the pathogen-host interactomes of Hepatitis C Virus/Human (2008), Epstein Barr virus/Human (2008), Influenza virus/Human (2009) were delineated through HTP to identify essential molecular components for pathogens and for their host's immune system.

Predicted Interactomes

As described above, PPIs and thus whole interactomes can be predicted. While the reliability of these predictions is debatable, they are providing hypotheses that can be tested experimentally. Interactomes have been predicted for a number of species, e.g.

- Human (Homo sapiens).
- Rice (Oryza sativa).
- Xanthomonas oryzae.
- Arabidopsis thaliana.
- Tomato.
- Brassica rapa.
- Maize, corn (Zea mays).
- Populus trichocarpa.

Network Properties

Protein interaction networks can be analyzed with the same tool as other networks. In fact, they share many properties with biological or social networks. Some of the main characteristics are as follows.

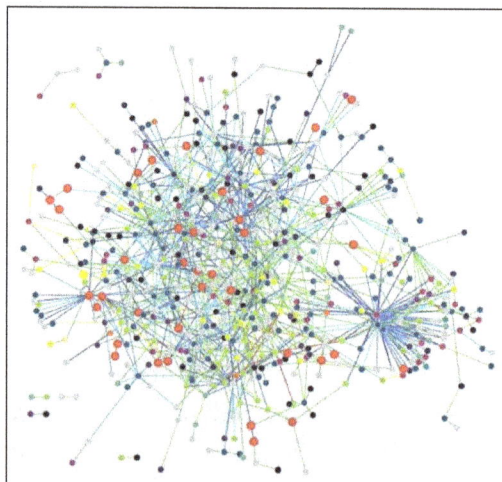

The Treponema pallidum protein interactome.

Degree Distribution

The degree distribution describes the number of proteins that have a certain number of connections. Most protein interaction networks show a scale-free (power law) degree distribution where the connectivity distribution $P(k) \sim k^{-\gamma}$ with k being the degree. This relationship can also be seen as a straight line on a log-log plot since, the above equation is equal to log(P(k)) ~ —y•log(k). One characteristic of such distributions is that there are many proteins with few interactions and few proteins that have many interactions, the latter being called "hubs".

Hubs

Highly connected nodes (proteins) are called hubs. Han et al. have coined the term "party hub" for hubs whose expression is correlated with its interaction partners. Party hubs also connect proteins within functional modules such as protein complexes. In contrast, "date hubs" do not exhibit such a correlation and appear to connect different functional modules. Party hubs are found predominantly in AP/MS data sets, whereas date hubs are found predominantly in binary interactome network maps. Note that the validity of the date hub/party hub distinction was disputed. Party hubs generally consist of multi-interface proteins whereas date hubs are more frequently single-interaction interface proteins. Consistent with a role for date-hubs in connecting different processes, in yeast the number of binary interactions of a given protein is correlated to the number of phenotypes observed for the corresponding mutant gene in different physiological conditions.

Modules

Nodes involved in the same biochemical process are highly interconnected.

Evolution

The evolution of interactome complexity is delineated in a study published in *Nature*. In this study it is first noted that the boundaries between prokaryotes, unicellular eukaryotes and multicellular eukaryotes are accompanied by orders-of-magnitude reductions in effective population size, with concurrent amplifications of the effects of random genetic drift. The resultant decline in the efficiency of selection seems to be sufficient to influence a wide range of attributes at the genomic level in a nonadaptive manner. The Nature study shows that the variation in the power of random genetic drift is also capable of influencing phylogenetic diversity at the subcellular and cellular levels. Thus, population size would have to be considered as a potential determinant of the mechanistic pathways underlying long-term phenotypic evolution. In the study it is further shown that a phylogenetically broad inverse relation exists between the power of drift and the structural integrity of protein subunits. Thus, the accumulation of mildly deleterious mutations in populations of small size induces secondary selection for protein–protein interactions that stabilize key gene functions, mitigating the structural degradation promoted by inefficient selection. By this means, the complex protein architectures and interactions essential to the genesis of phenotypic diversity may initially emerge by non-adaptive mechanisms.

Structural Biology

Structural biology is the study of how biological molecules are built. Using a variety of imaging techniques, scientists view molecules in three dimensions to see how they are assembled, how they function, and how they interact. That has helped researchers understand how the thousands of different molecules in each of our cells work together to keep us healthy. Structural studies have also shown how misshapen molecules make us sick, and as a result, these studies have prompted new treatments for many diseases.

Molecules

Molecules are groups of two or more atoms held together by chemical bonds. Molecules include DNA, RNA, proteins, carbohydrates (sugars), and lipids (fats). Structural biologists are particularly interested in proteins because they do so much of the work in the body. Increasingly, biologists are investigating large molecules made up of combinations of RNA and proteins, called RNA-protein complexes.

Proteins

Proteins are molecules that contribute to virtually every activity in the body. They form hair and fingernails, carry oxygen in the blood, allow muscles to move, and much more.

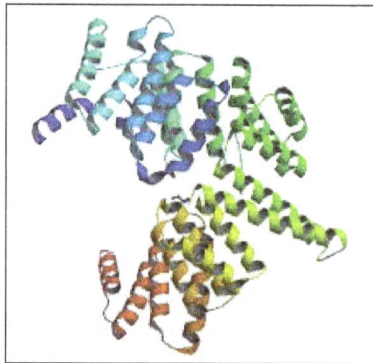

Protein molecules are made up of amino acids hooked together like beads on a string. To become active, proteins must twist and fold into their final, or "native," configuration.

What are Proteins made of?

Proteins are made of long strands of small molecules called amino acids. There are 20 amino acids found in nature. Each protein contains a unique combination of a few dozen to many thousands of amino acids. Some proteins consist of multiple amino acid strands wound together.

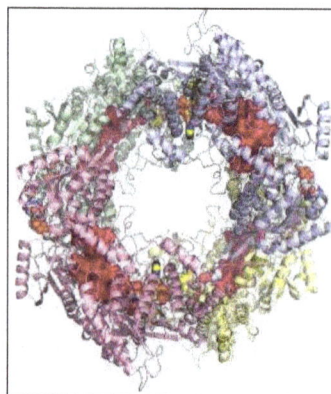

A mix of alpha helices (red; curled ribbon forming a six-sided star) and beta sheets (blue; thinner, tangled strands).

How does a Protein Get its Shape?

Even though proteins are strings of amino acids, they do not remain in a straight line. The strands twist, bend, and fold into specific shapes. The way they fold depends in part on the way the amino acids interact with each other. Some sections of proteins form standard "motifs": corkscrew-like coils called alpha helices and flat sections called beta sheets.

Researchers can easily determine a protein's amino acid sequence. The trick has been figuring out how and why the proteins fold. Scientists are beginning to solve that puzzle with research into how amino acids interact, and they are using powerful new computer programs that help predict protein motifs. Researchers are even starting to design their own brand-new proteins that perform specific jobs. This new work helps scientists understand not only how proteins fold but also how they misfold and malfunction in diseases such as Alzheimer's and cystic fibrosis. Knowing more about these processes might allow researchers to design new treatments.

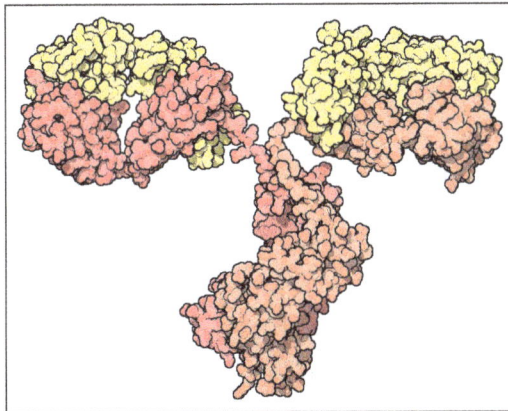

An image of an antibody, an immune system protein that rids the body of foreign material, including bacteria.

Why does a Protein's Shape Matter?

A protein's structure allows it to perform its job. For instance, antibodies are shaped like a Y. This helps these immune-system proteins bind to foreign molecules such as bacteria or viruses with one end while recruiting other immune-system proteins with the other. DNA polymerase III is donut-shaped. This helps it form a ring around DNA as it copies its genetic information. And proteins called enzymes have grooves and pockets that help them hold onto other molecules to speed chemical reactions. Misfolded, or misshapen, proteins can cause diseases. They often stop working properly and can build up in tissues. Alzheimer's disease, Parkinson's disease, and cystic fibrosis are examples of diseases caused by misfolded proteins.

What Kinds of Proteins are there?

There are many different kinds of proteins. For example, many proteins are enzymes that aid biochemical reactions. Others have shapes or specific functions that help cells hold their shape and move, as described in the previous answer. Another type, called transporter proteins, are embedded in the cell's outer membrane and form channels that help vital substances such as sodium or potassium pass into or out of the cell.

Protein Structures to Develop New Drugs

Drugs typically work by either blocking or supporting the activity of specific proteins in the body. Using an approach called structure-based drug design, scientists can make a template for a protein and use that blueprint for creating new medicines. They start with a computerized model of the protein structure they're interested in studying. For example, the computer model would allow researchers to examine how two proteins work together. Then, if scientists want to turn off one protein, they would try to design a molecule that would block or alter that interaction.

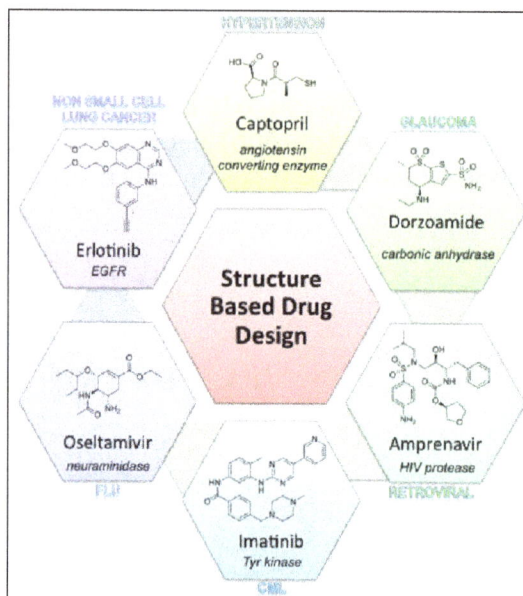

Knowing that HIV protease—an enzyme that breaks down HIV—has two symmetrical halves, pharmaceutical researchers initially attempted to block the enzyme with symmetrical, naturally occurring small molecules. They made these by chopping in half molecules of the natural substrate, then making a new molecule by fusing together two identical halves of the natural substrate.

Structure-based Drug Design

Researchers used structure-based drug design to develop some anti-HIV drugs. HIV protease is an enzyme that keeps the virus alive. Knowing its structure allowed researchers to determine the kinds of molecules that could stop HIV protease from working. Scientists used computer models to fine tune molecules that could halt virus production. This work led to medicines called protease inhibitors.

Protein Structures

Researchers use several imaging techniques to determine the structure of proteins and other complex molecules. Cryo-electron microscopy (cryo-EM) allows scientists to "see" individual proteins as well as larger structures such as molecular complexes (groups of proteins that combine and function as a unit), viruses, or organelles (specialized structures within the cell that perform specific functions). X-ray crystallography and nuclear magnetic resonance (NMR) spectroscopy also make it possible for researchers to view proteins. To date, researchers have used these techniques to unravel the structure of more than 122,000 proteins.

Cryo-EM

In cryo-EM, researchers rapidly freeze a cell, virus, molecular complex, or other structure so that water molecules do not have time to form crystals. This preserves the sample in its natural state. Scientists use an electron microscope to blast the frozen sample with an electron beam. This creates a two-dimensional projection of the sample on a digital detector. By creating hundreds of projections of the sample from many different angles and then taking the average of these angles, scientists generate a three-dimensional model of its structure. Recent advances in cryo-EM provide highly detailed images of proteins and other biological structures, including larger structures such as RNA-protein complexes.

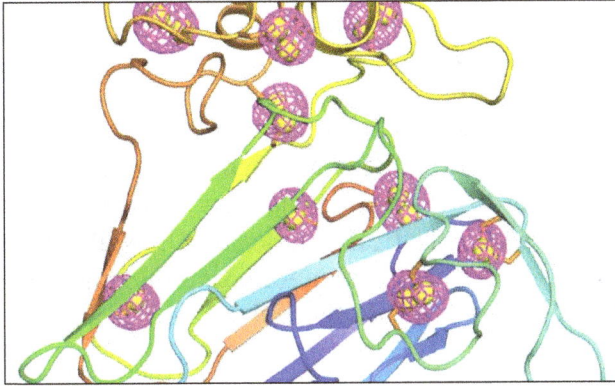

Structural biologists create crystals of proteins, shown here, as step in X-ray crystallography, which can reveal detailed, three-dimensional protein structures.

X-ray Crystallography

X-ray crystallography shoots a beam of X-rays through a tiny solid crystal made up of trillions of identical protein molecules. The crystal scatters the X-rays onto an electronic detector, similar to the way images are captured in a digital camera. A computer gauges the intensities of the scattered X-rays to assign a position to each atom in the crystallized molecule. The result is a three-dimensional digital image. This method has been used to determine more than 85 percent of known protein structures.

NMR Spectroscopy

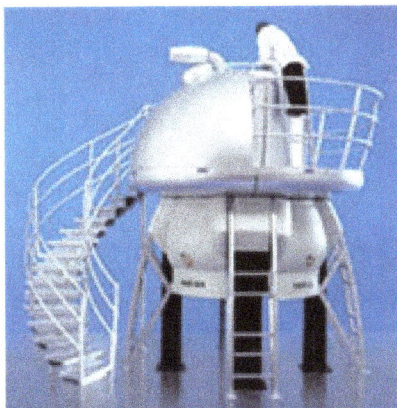

This is a machine used to do NMR spectroscopy. Most of these machines use magnets that are 500 megahertz to 900 megahertz. This magnet is 900 megahertz.

NMR spectroscopy works using the natural magnets—the nuclei of certain atoms—inside proteins. Those natural cellular magnets interact with a big magnet inside the NMR machine. The big magnet forces the protein's magnets to line up. Researchers then blast the sample with a series of split-second radio-wave pulses and observe how the protein's magnets respond. Scientists use several sets of these NMR blasts and combine the data to get a more complete picture of the protein. Although X-ray crystallography can examine larger proteins than NMR is able to do, NMR technology can study proteins immersed in liquid solutions. In contrast, X-ray crystallography requires that proteins be organized into crystals.

Can Scientists view how Proteins Act?

New technology is beginning to allow researchers to progress from creating static pictures of proteins and other molecules to making movies of their actions.

Images provide snapshots of what these cellular elements are doing at specific points in time. Although they supply valuable information, these still pictures don't capture how proteins and other molecules inside cells are constantly moving and changing, folding and unfolding as they interact. Understanding this dynamic system is critical to unlocking how life works. In addition, there is a whole class of proteins, called intrinsically disordered proteins, that do not hold a specific shape. Their shape adapts to what's going on inside the cell, making it nearly impossible to take a still picture of them.

Scientists are now using powerful computer models to make molecular movies so they can see the full range of proteins in live action. By feeding in information from various imaging techniques about how amino acids and other building blocks interact, scientists can create moving pictures. Such movies help researchers understand how proteins work in their natural state and allow them to design highly specific drugs.

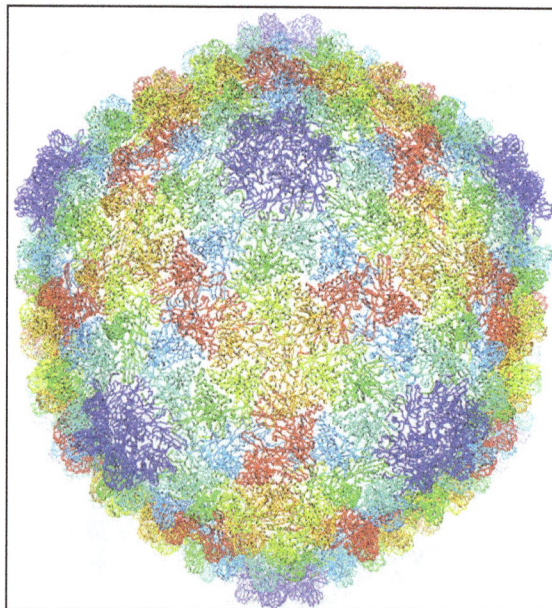

An atomic-scale model of a virus that
infects the Salmonella bacterium.

Future of Structural Biology

Researchers on the frontier of structural biology are merging all of the imaging techniques—X-ray crystallography, NMR, and cryo-EM. This allows them to create a more precise map of what proteins and other molecules look like and how they interact. Scientists can create a single image that zooms in to see specific proteins and also zooms out to see how they interact within the larger cellular structure. In addition to combining existing techniques, scientists are developing ever more powerful methods. For example, new X-ray lasers allow insights into processes that occur in less than one tenth of a trillion of a second, much faster timescales than that captured by other sources of X-rays.

Scientists use super-efficient methods to determine protein structures more quickly than ever before. They also use sophisticated techniques to predict three-dimensional structures of proteins. And they use high-powered computer models to design and create new proteins not found in nature that have useful functions, such as discovering and combating disease. This work will continue to increase our understanding of the diverse roles molecules play in biology and to spur advances in medicine.

References

- Tomar, Rukam (2010). Molecular Markers and Plant Biotechnology. Pitman Pura, New Delhi: New India Publishing Agency. P. 188. ISBN 978-93-80235-25-7

- Molecular-biology, science: britannica.com, Retrieved 19 April, 2019

- Watson JD (2007). Recombinant DNA: genes and genomes: a short course. San Francisco: W.H. Freeman. ISBN 978-0-7167-2866-5

- Pfeifer A, Verma IM (2001). "Gene therapy: promises and problems". Annual Review of Genomics and Human Genetics. 2: 177–211. Doi:10.1146/annurev.genom.2.1.177. PMID 11701648

- Ninfa, Alexander; Ballou, David; Benore, Marilee (2009). Fundamental Laboratory Approaches for Biochemistry and Biotechnology. United States: Wiley. Pp. 408–410. ISBN 978-0470087664

- Hoffmann, R; Krallinger, M; Andres, E; Tamames, J; Blaschke, C; Valencia, A (2005). "Text mining for metabolic pathways, signaling cascades, and protein networks". Science Signaling. 2005 (283): pe21. Doi:10.1126/stke.2832005pe21. PMID 15886388

- Factsheet-structural biology, education: nigms.nih.gov, Retrieved 5 February, 2019

5
Systems Biology

The interdisciplinary field of study which focuses on complex interactions within biological systems through a holistic approach towards biological research is known as systems biology. It is involved in the computational and mathematical analysis and modeling of complex biological systems. All the diverse principles and models related to systems biology have been carefully analyzed in this chapter.

Systems biology is the study of the interactions and behavior of the components of biological entities, including molecules, cells, organs, and organisms.

The organization and integration of biological systems has long been of interest to scientists. Systems biology as a formal, organized field of study, however, emerged from the genomics revolution, which was catalyzed by the Human Genome Project and the availability to biologists of the DNA sequences of the genomes of humans and many other organisms. The establishment of the field was also influenced heavily by the general recognition that organisms, cells, and other biological entities have an inherently high degree of complexity. Two dominant themes of modern biology are rooted in that new outlook: first, the view that biology is fundamentally an informational science—biological systems, cells, and organisms store and transfer information as their most-fundamental processes—and second, the emergence of new technologies and approaches for studying biological complexity.

Biological organisms are very complex, and their many parts interact in numerous ways. Thus, they can be considered generally as integrated systems. However, whereas an integrated complex system such as that of a modern airliner can be understood from its engineering design and detailed plans, attempting to understand the integrated system that is a biological organism is far more difficult, primarily because the number and strengths of interactions in the system are great and they must all be inferred after the fact from the system's behavior. In the same manner, the blueprint for its design must be inferred from its genetic material. That "integrated systems" point of view and all the associated approaches for the investigation of biological cells and organisms are collectively called systems biology.

Complexity and Emergent Properties

Many of the most-critical aspects of how a cell works result from the collective behavior of many molecular parts, all acting together. Those collective properties—often called "emergent properties"—are critical attributes of biological systems, as understanding the individual parts alone is insufficient to understand or predict system behavior. Thus, emergent properties necessarily come from the interactions of the parts of the larger system. As an example, a memory that is stored in the human brain is an emergent property because it cannot be understood as a property of a single neuron or even many neurons considered one at a time. Rather, it is a collective property of a large number of neurons acting together.

One of the most-important aspects of the individual molecular parts and the complex things they constitute is the information that the parts contain and transmit. In biology information in molecular structures—the chemical properties of molecules that enable them to recognize and bind to one another—is central to the function of all processes. Such information provides a framework for understanding biological systems, the significance of which was captured insightfully by American theoretical physical chemist Linus Pauling and French biologist Emil Zuckerkandl, who stated in a joint paper, "Life is a relationship among molecules and not a property of any one molecule." In other words, life is defined in terms of interactions, relationships, and collective properties of many molecular systems and their parts.

The central argument concerning information in biology can be seen by considering the heredity of information, or the passing on of information from one generation to the next. For a given species, the information in its genome must persist through reproduction in order to guarantee the species' survival. DNA is passed on faithfully, enabling a species' genetic information to endure and, over time, to be acted on by evolutionary forces. The information that exists in living things today has accumulated and has been shaped over the course of more than 3.4 billion years. As a result, focusing on the molecular information in biological systems provides a useful vantage point for understanding how living systems work.

That the emergent properties derived from the collective function of many parts are the key properties of biological systems has been known since at least the first half of the 20th century. They have been considered extensively in cell biology, physiology, developmental biology, and ecology. In ecology, for example, debate regarding the importance of complexity in ecological systems and the relationship between complexity and ecological stability began in the 1950s. Since then, scientists have realized that complexity is a general property of biology, and technologies and methods to understand parts and their interactive behaviors at the molecular level have been developed. Quantitative change in biology, based on biological data and experimental methods, has precipitated profound qualitative change in how biological systems are viewed, analyzed, and understood. The repercussions of that change have been immense, resulting in shifts in how research is carried out and in how biology is understood.

A comparison with systems engineering can provide useful insight into the nature of systems biology. When engineers design systems, they explore known components that can be put together in such a way as to create a system that behaves in a prescribed fashion, according to the design specifications. When biologists look at a system, on the other hand, their initial tasks are to identify the components and to understand the properties of individual components. They then attempt to identify how interactions between the components ultimately create the system's observable biological behaviors. The process is more closely aligned with the notion of "systems reverse engineering" than it is with systems design engineering.

The Human Genome Project contributed broadly to that revolution in biology in at least three different ways: (1) by acquiring the genetics "parts list" of all genes in the human genome; (2) by catalyzing the development of high-throughput technology platforms for generating large data sets for DNA, RNA, and proteins; and (3) by inspiring and contributing to the development of the computational and mathematical tools needed for analyzing and understanding large data sets. The project, it could be argued, was the final catalyst that brought about the shift to the systems point of view in biology.

Ukrainian American geneticist and evolutionist Theodosius Dobzhansky noted in the 20th century that "nothing makes sense in biology except in the light of evolution." Understanding evolution is essential to systems biology, but understanding where the information in the system came from and how it became complex also provides a focus for evolutionary thought. In a series of evolution-ary transitions, biological systems have acquired remarkable mechanisms for storing, handling, and deploying information in the living world. The fundamental parts for storage and transmis-sion are RNA and DNA. A striking insight that emerged from the study of those parts is that the evolution of developmental and physiological systems has involved basic components of gene-reg-ulatory networks, including transcription factors (a class of proteins that regulate gene expression) and transcription factor-binding cis-regulatory elements in DNA. Gene-regulatory networks are coupled in turn to other networks that have profound effects on the function of systems and there-by determine evolutionary possibilities.

Networks and Information

Engineers and mathematicians have provided valuable insights into the nature of information, par-ticularly related to communications, and biologists have adapted some of those insights to the study of biological systems. A significant area of research in biology centres on the question of whether higher-order biological processes can be represented from an information perspective. The concep-tual tools for looking at biological phenomena are based on mathematical ideas about information and computing, but significant further development is required before a satisfactory theoretical ba-sis is realized. For example, a key aspect of describing and measuring biological information is the context in which the information operates, which has been difficult to represent in a clear and useful way. An example of the type of challenge that researchers face is the process of gene expression, which involves the production of a specific protein molecule from genetic information. A number of factors impinge on the expression of any one gene—from the type of cell involved to the external sig-nals received by and the metabolic state of the cell to preexisting states of gene expression. Efforts to understand those factors form a major area of research in modern biology.

Although some small networks, such as certain metabolic networks in bacteria or yeast, are relatively well characterized, more-complex networks, such as developmental networks, remain only partially understood. Mathematical concepts relevant to the study of both types of networks have been devel-oped and implemented. Still, few biological systems have been characterized sufficiently to enable researchers to model them as networks. Examples include the lactose- and galactose-utilization sys-tems in certain bacteria, such as Escherichia coli and Streptococcus. However, wider interactions of those networks are comparatively less well understood. The early embryonic development of the sea urchin is another system that has been effectively modeled. Models offer unique insight into biolog-ical development and physiology, and scientists have envisioned a future when models will become available for most biological systems. Indeed, quantitative models could ultimately come to embody hypotheses about the structure and function of any biological system in question.

Ideas about biological systems have the potential to transform agriculture, animal husbandry, nutrition, energy, and other industries and fields of research. By the early 21st century, the practice of medicine had already begun to move from reactive treatment of patients' symptoms to proactive and personalized care because of improved understanding of the functions of the complex systems that are human cells and organs.

Biological Processes

Biological process are the processes vital for a living organism to live, and that shape its capacities for interacting with its environment. Biological processes are made up of many chemical reactions or other events that are involved in the persistence and transformation of life forms. Metabolism and homeostasis are examples.

Regulation of biological processes occurs when any process is modulated in its frequency, rate or extent. Biological processes are regulated by many means; examples include the control of gene expression, protein modification or interaction with a protein or substrate molecule.

- Homeostasis: Regulation of the internal environment to maintain a constant state; for example, sweating to reduce temperature.

- Organization: Being structurally composed of one or more cells – the basic units of life.

- Metabolism: Transformation of energy by converting chemicals and energy into cellular components (anabolism) and decomposing organic matter (catabolism). Living things require energy to maintain internal organization (homeostasis) and to produce the other phenomena associated with life.

- Growth: Maintenance of a higher rate of anabolism than catabolism. A growing organism increases in size in all of its parts, rather than simply accumulating matter.

- Adaptation: The ability to change over time in response to the environment. This ability is fundamental to the process of evolution and is determined by the organism's heredity, diet, and external factors.

- Response to stimuli: A response can take many forms, from the contraction of a unicellular organism to external chemicals, to complex reactions involving all the senses of multicellular organisms. A response is often expressed by motion; for example, the leaves of a plant turning toward the sun (phototropism), and chemotaxis.

- Reproduction: The ability to produce new individual organisms, either asexually from a single parent organism or sexually from two parent organisms.

- Interaction between organisms: The processes by which an organism has an observable effect on another organism of the same or different species.

- Also: Cellular differentiation, fermentation, fertilisation, germination, tropism, hybridisation, metamorphosis, morphogenesis, photosynthesis, transpiration.

Biosimulation

Biosimulation is computer-aided mathematical simulation of biological processes and systems and thus is an integral part of systems biology. Due to the complexity of biological systems simplified models are often used, which should only be as complex as necessary.

The aim of biosimulations is model-based prediction of the behavior and the dynamics of biological systems e.g. the response of an organ or a single cell towards a chemical. However the quality of model-based predictions strongly depends on the quality of the model, which in turn is defined by the quality of the data and the profoundness of the knowledge.

Biosimulation is becoming increasingly important for drug development. Since on average only 11% of all drug candidates are approved, it is anticipated that biosimulation may be the tool to predict whether a candidate drug will fail in the development process e.g. in clinical trials due to adverse side effects, bad pharmacokinetics or even toxicity. The early prediction if a drug will fail in animals or humans would be a key to reduce both drug development costs and the amount of required animal experiments and clinical trials. The latter is also in line with the so-called "3Rs" which refer to the principle of reduction and replacement of animal experiments as well as to the refinement of the methodology in cases where animal tests are still necessary. In a future scenario, biosimulation would change the way substances are tested, in which in vivo and in vitro tests are substituted by tests in silico.

Due to the importance of biosimulation in drug development a number of research projects exist which aim for simulating metabolism, toxicity, pharmacodynamic and pharmacokinetics of a drug candidate. Some of the research projects are listed below:

- BioSim project; funded by the 6th frame program of the European Union.
- NSR Physiome Project.
- Hepatosys.

Moreover, a few software tools already exist, which aim for predicting the toxicity of a substance or even try to simulate the virtual patient (Entelos). A few of these software tools are listed below:

- COPASI.
- Metabolism PhysioLab (Entelos).
- GastroPlus and ADMEPredictor (Simulations-Plus).
- Certara (previously Pharsight)'s computer assisted trial design.
- RHEDDOS (Rhenovia Pharma SAS).
- VirtualToxLab.
- Derek (Lhasa Limited).
- Simcyp.
- DS TOPKAT (accelrys).
- ADME Workbench.

Biological Systems

A biological system is a complex network of biologically relevant entities. Biological organization spans several scales and are determined based different structures depending on what the system is. Examples of biological systems at the macro scale are populations of organisms. On the organ and tissue scale in mammals and other animals, examples include the circulatory system, the respiratory system, and the nervous system. On the micro to the nanoscopic scale, examples of biological systems are cells, organelles, macromolecular complexes and regulatory pathways. A biological system is not to be confused with a living system, such as a living organism.

Organ and Tissue Systems

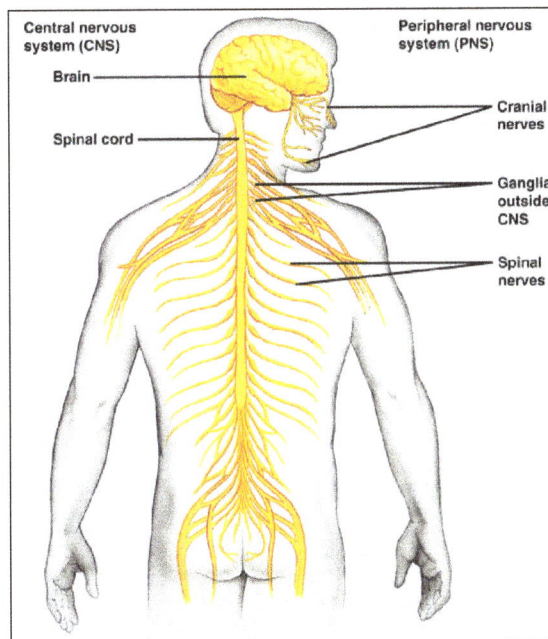

An example of a system: The brain, the cerebellum, the spinal cord, and the nerves are the four basic components of the nervous system.

These specific systems are widely studied in human anatomy and are also present in many other animals.

- Respiratory system: The organs used for breathing, the pharynx, larynx, bronchi, lungs and diaphragm.

- Digestive system: Digestion and processing food with salivary glands, oesophagus, stomach, liver, gallbladder, pancreas, intestines, rectum and anus.

- Cardiovascular system (heart and circulatory system): Pumping and channeling blood to and from the body and lungs with heart, blood and blood vessels.

- Urinary system: Kidneys, ureters, bladder and urethra involved in fluid balance, electrolyte balance and excretion of urine.

- Integumentary system: Skin, hair, fat, and nails.

- Skeletal system: Structural support and protection with bones, cartilage, ligaments and tendons.

- Endocrine system: Communication within the body using hormones made by endocrine glands such as the hypothalamus, pituitary gland, pineal body or pineal gland, thyroid, parathyroid and adrenals, i.e., adrenal glands.

- Lymphatic system: Structures involved in the transfer of lymph between tissues and the blood stream; includes the lymph and the nodes and vessels. The lymphatic system includes functions including immune responses and development of antibodies.

   ○ Immune system: Protects the organism from foreign bodies.

- Nervous system: Collecting, transferring and processing information with brain, spinal cord, peripheral nervous system and sense organs.

   ○ Sensory systems: Visual system, auditory system, olfactory system, gustatory system, somatosensory system, vestibular system.

- Muscular system: Allows for manipulation of the environment, provides locomotion, maintains posture, and produces heat. Includes skeletal muscles, smooth muscles and cardiac muscle.

- Reproductive system: The sex organs, such as ovaries, fallopian tubes, uterus, vagina, mammary glands, testes, vas deferens, seminal vesicles and prostate.

Cellular Organelle Systems

The exact components of a cell are determined by whether the cell is a eukaryote or prokaryote.

- Nucleus: Storage of genetic material; control center of the cell.

- Cytosol: Component of the cytoplasm consisting of jelly-like fluid in which organelles are suspended within.

- Cell membrane (plasma membrane):

- Endoplasmic reticulum: Outer part of the nuclear envelope forming a continuous channel used for transportation; consists of the rough endoplasmic reticulum and the smooth endoplasmic reticulum.

   ○ Rough endoplasmic reticulum (RER): Considered "rough" due to the ribosomes attached to the channeling; made up of cisternae that allow for protein production.

   ○ Smooth endoplasmic reticulum (SER): Storage and synthesis of lipids and steroid hormones as well as detoxification.

- Ribosome: Site of biological protein synthesis essential for internal activity and cannot be reproduced in other organs.

- Mitochondrion (mitochondria): Powerhouse of the cell; site of cellular respiration producing ATP (adenosine triphosphate).

- Lysosome: Center of breakdown for unwanted/unneeded material within the cell.

- Peroxisome: Breaks down toxic materials from the contained digestive enzymes such as H_2O_2(hydrogen peroxide).

- Golgi apparatus (eukaryotic only): Folded network involved in modification, transport, and secretion.

- Chloroplast (prokaryotic only): Site of photosynthesis; storage of chlorophyll.

Biological Network

Networks are widely used in many branches of biology as a convenient representation of patterns of interaction between appropriate biological elements. These biological networks include biochemical networks, neural networks, and ecological networks.

Biochemical Networks

Biochemical networks represent the molecular-level patterns of interaction and mechanisms of control in the biological cell. The principal types of these networks are metabolic networks, protein-protein networks, and genetic regulatory networks.

Metabolic Networks

Metabolism is the chemical process by which cells break down food and nutrients into usable building blocks and then reassemble those building block to form the biological molecules the cell needs to complete its other tasks. Typically this breakdown and reassembly involves chains or pathways, sets of successive chemical reactions that convert initial inputs into useful end products by a series of steps. The complete set of all reactions in all pathways forms the metabolic network.

The vertices in a metabolic network are chemicals produced and consumed by the reactions. These chemicals are known as metabolites. These are small molecules like carbohydrates, lipids, as well as amino acids and nucleotides. The metabolites consumed are called the substrates of the reaction, while those produced are called the products.

Most metabolic reactions do not occur spontaneously, or do so only at a very low rate. To make reactions occur at a usable rate, the cell employs an array of chemical catalysts, referred to as enzymes. Enzymes are not consumed in the reactions they catalyze. By increasing or decreasing the concentration of the enzyme that catalyzes a particular reaction, the cell can turn that reaction on or off, or moderate its speed. Enzymes tend to be highly specific to the reaction they catalyze.

The most correct representation of a metabolic network is as a bipartite graph. The two types of vertices represent metabolites and metabolic reactions, with edges joining each metabolite to the reaction in which it participates. The edges are directed, since some metabolites (the substrates) go into the reaction and some (the products) come out of it. Enzymes can be incorporated by adding a third class of vertex to represent them, with undirected edges connecting them to the reactions they catalyze. The resulting graph is a mixed (directed and undirected) tripartite network.

Nevertheless, the most common representations of metabolic network project the bipartite network onto metabolite vertices. In one approach, the vertices in the network represent metabolites and there is an undirected edge between any two metabolites that participate in the same reaction, either as substrates or as products. Clearly this projection loses much of the information contained in the bipartite network. Another more informative approach is to represent the network as a directed graph with a single type of vertex representing metabolites and a directed edge from one metabolite to another if there is a reaction in which the first metabolite appears as a substrate and the second as a product. This representation is more reach in terms of information with respect to the undirected one but still loses the association of metabolites with reactions.

Protein-protein Interaction Networks

Proteins do interact with one another and with other biomolecules, both large and small, but the interactions are not purely chemical. Proteins are long-chain molecules formed by the concatenation of a series of basic units called amino acids. Once created, a protein does not stay in a loose chain-like form, but folds on itself in a folded form whose shape depends on the amino acid sequence. The folded form dictates the physical interaction it can have with other molecules. Hence, the primary mode of protein-protein interaction is physical rather than chemical, their complicated folded shapes interlocking to create so-called protein complexes but without the exchange of particles that defines chemical reactions.

In a protein-protein interaction network the vertices are proteins and two vertices are connected by an undirected edge if the corresponding protein interact. However, this representation omits useful information. Interactions that involve three or more proteins are represented by multiple edges, and there is no way to tell from the network itself that such edges represent aspects of the same interaction. This problem could be addressed by adopting a bipartite representation, with proteins and interactions as different types of vertex, and undirected edges connection proteins to the interactions in which they participate.

Genetic Regulatory Networks

The small molecules needed by biological organisms, such as sugars and fats, are manufactured in the cell by the chemical reactions of metabolism. Proteins, however, which are much larger molecules, are manufactured in a different manner, following recipes recorded in the cell's genetic material, DNA.

Proteins are long-chain molecules formed by the concatenation of amino acids. The protein amino acid sequence is determined by a corresponding sequence stored in the DNA of the cell in which the protein is synthesized. This is the primary function of DNA, to act as an information storage medium containing the sequences of proteins needed by the cell. DNA is itself made up of units called nucleotides, of which there are four distinct species, denoted A, C, G, and T. The amino acids in proteins are encoded in DNA as trios of consecutive nucleotides called codons, such as ACG and TTT, and a succession of codons spells out the complete sequence of amino acids in a protein. A single strand of DNA can code for many proteins, and two special codons, called the start and end codons, signal the beginning and end of the sequence coding for a protein. The DNA code for a single protein, from start to stop codon, is called a gene.

Proteins are created in the cell by a mechanism that operates in two stages. In the first stage, known as transcription, an enzyme makes a copy of the coding sequence of a single gene. The copy is made of RNA, another information-bearing similar to DNA. In the second stage, called translation, the protein is assembled step by step from the RNA sequence. In the jargon of molecular biology, one says that the gene has been expressed.

The cell does not, in general, need to produce at all times every possible protein for which it contains a gene. Individual proteins serve specific purposes, such as catalyzing metabolic reactions, and it is important for the cell to respond to its environment by turning on or off the production of individual proteins. It does this by the use of transcription factors, which are themselves proteins. The presence in the cell of the transcription factor for the gene turns on or enhances the expression of that gene, or inhibits it, depending on the type of transcription factor (promoting and inhibiting).

Here comes the network. Being proteins, transcription factors are themselves produced by transcription from genes, and the transcription process is regulated by other transcription factors, which again are proteins, and so forth. The complete set of such interactions forms a genetic regulatory network. The vertices in this network are proteins, or equivalently the genes that code for them and a directed edge from gene A to gene B indicates that A regulates the expression of B. One can distinguish between promoting and inhibiting transcription factors, giving the network two distinct types of edges.

Neural Networks

One of the main functions of the brain is to process information and the primary information processing element is the neuron, a specialized brain cell that combines several inputs to generate a single output.

A typical neuron consists of a cell body or soma, along with a number of protruding tentacles, called dendrites, which are input wires for carrying signals in the cell. Most neurons have only one output, called the axon, which is typically longer than the dendrites. It usually branches near its end into axon terminals to allow the output of the cell to feed the input of several others. There is a small gap, called synapse, between terminal and dendrite across which the output signal of the first (presynaptic) neuron must be conveyed to reach the second (postsynaptic) neuron.

The actual signals that travel within neurons are electrochemical in nature. They consists in travelling waves of electrical voltage created by the motion of positively charged ions in and out of the cell. These waves are called action potentials. When an action potential reaches a synapse, it cannot cross the gap between the axon terminal and the opposing dendrite and the signal is instead transmitted chemically. The arrival of the action potential stimulates the production of a chemical neuro-transmitter by the terminal, which diffuses across the gap and is detected by receptor molecules on the dendrite at the other side. This in turn causes ions to move in and out the dendrite, changing its voltage.

These voltage changes, however, do not yet give rise to another travelling wave. The soma of the postsynamptic neuron sums the inputs from its dendrites and as a result may or may not send an output signal down its own axon. Thus the neuron acts as a gate that aggregates the signals at its inputs and only fires when enough inputs are excited. Furthermore, inputs can also be inhibiting; signals received at inhibiting inputs make the receiving neuron less likely to fire.

Current science cannot tell us exactly how the brain performs the more sophisticated cognitive tasks that allow animals to survive, but it is known that the brain constantly changes the pattern of wiring between neurons in response to inputs and experience, and it is presumed that this pattern - the neural network - holds much of the secret.

At the simplest level, a neural network can be represented as a set of vertices, the neurons, connected by two types of directed edges, one for excitatory inputs and one for inhibiting inputs. In practice, neurons are not all the same. This variation can be encoded in our network representation by different types of vertices.

Ecological Networks

Ecological networks are networks of ecological interactions between species. Species in an ecosystem can interact in different ways: they can eat one another, they can parasitize one another, or they can have any of a variety of mutually advantageous interactions, such as pollination or seed dispersal.

Food webs are the most studied ecological networks. The biological organisms on our planet can be divided into ecosystems, groups of organisms that interact with one another and with elements of their environment. A food web is a directed network that represents which species prey on which others in a given ecosystem. The vertices of the network correspond to species and the directed edges to predator-prey interactions. In fact, ecologists conventionally draw edges in the opposite direction, from prey to predator, that is, in the direction of energy flow.

Food webs are approximately directed acyclic graphs (DAGs). The acyclic nature of food webs indicate that there is an intrinsic hierarchy among the species in ecosystems. Those higher up the hierarchy prey on those lower down, but not vice versa, although there are some counterexamples. A species's position in this hierarchy is called by ecologists its trophic level. This is the rank of the species' vertex on the acyclic food graph, that is the length of the longest path leading to the vertex representing the species. Those species in lower trophic levels tend to be smaller and more abundant, while those in higher trophic positions are usually larger-bodied and less numerous predators.

Community food webs are complete food webs for an entire ecosystem. Source food webs and sink food webs are sub-networks of community food webs that focus on species connected to a specific prey or predator. In a source food web, one records all species that derive energy from a particular source species. A sink food web, on the other hand, specifies through which species the energy is, directly or indirectly, consumed by a given sink species. In some cases attempts have been made to measure not merely the presence of interactions between species but also their strength, for instance as the fraction of energy that a species derives from each of its preys. The result is a weighted directed network that sheds more light on the flow of energy through an ecosystem.

Other ecological networks include host-parasite networks and mutualistic networks. Host-parasite networks are networks of parasitic relationships between organisms. Parasites tend to be smaller-bodied then their hosts and parasites can live off their hosts without killing them. Host-parasite networks are directed acyclic networks. Mutualistic networks are networks of mutually beneficial interactions between species. These include networks of plants and the animals (insects) that pollinate them, networks of plants and the animals (birds) that disperse their seeds, and networks of ant species and the plants that thy protect and eat. Mutualistic networks can be represented as undirected bipartite graphs.

Modelling Biological Systems

Modelling biological systems is a significant task of systems biology and mathematical biology. Computational systems biology aims to develop and use efficient algorithms, data structures, visualization and communication tools with the goal of computer modelling of biological systems. It involves the use of computer simulations of biological systems, including cellular subsystems (such as the networks of metabolites and enzymes which comprise metabolism, signal transduction pathways and gene regulatory networks), to both analyze and visualize the complex connections of these cellular processes.

Artificial life or virtual evolution attempts to understand evolutionary processes via the computer simulation of simple (artificial) life forms.

An unexpected emergent property of a complex system may be a result of the interplay of the cause-and-effect among simpler, integrated parts. Biological systems manifest many important examples of emergent properties in the complex interplay of components. Traditional study of biological systems requires reductive methods in which quantities of data are gathered by category, such as concentration over time in response to a certain stimulus. Computers are critical to analysis and modelling of these data. The goal is to create accurate real-time models of a system's response to environmental and internal stimuli, such as a model of a cancer cell in order to find weaknesses in its signalling pathways, or modelling of ion channel mutations to see effects on cardiomyocytes and in turn, the function of a beating heart.

Standards

By far the most widely accepted standard format for storing and exchanging models in the field is the Systems Biology Markup Language (SBML) The SBML.org website includes a guide to many important software packages used in computational systems biology. A large number of models encoded in SBML can be retrieved from BioModels. Other markup languages with different emphases include BioPAX and CellML.

Cellular Model

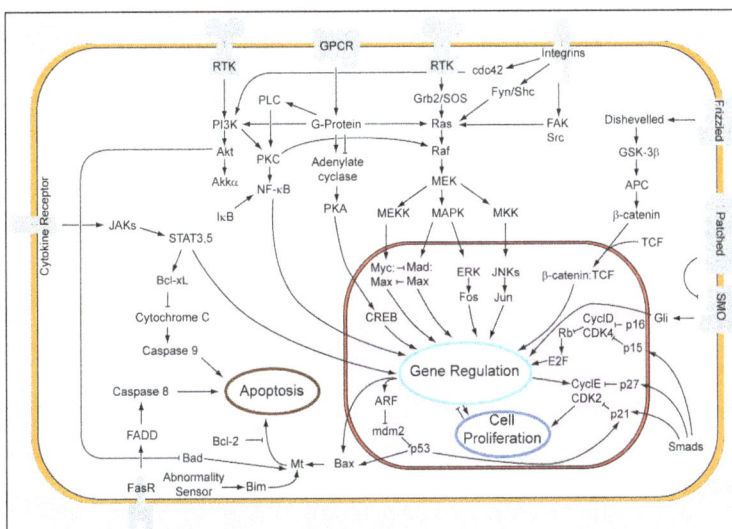

Part of the cell cycle.

Creating a cellular model has been a particularly challenging task of systems biology and mathematical biology. It involves the use of computer simulations of the many cellular subsystems such as the networks of metabolites and enzymes which comprise metabolism, signal transduction pathways and gene regulatory networks to both analyze and visualize the complex connections of these cellular processes.

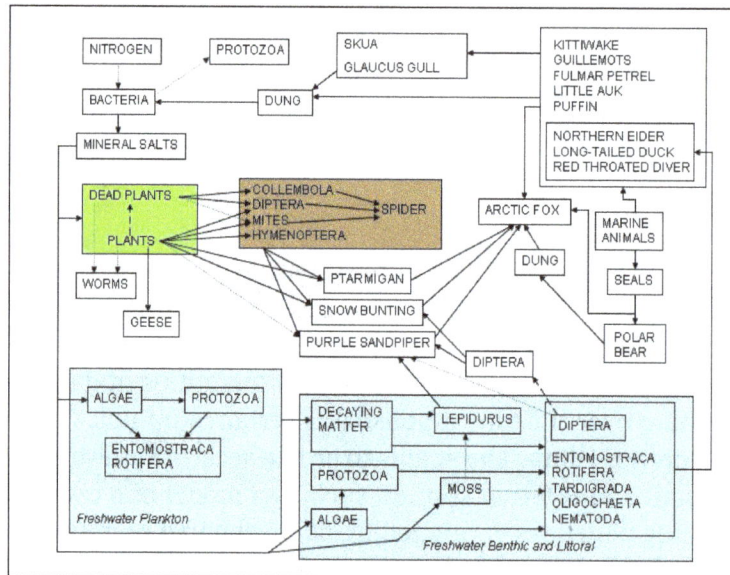

Summerhayes and Elton's 1923 food web of Bear Island.

The complex network of biochemical reaction/transport processes and their spatial organization make the development of a predictive model of a living cell a grand challenge for the 21st century, listed as such by the National Science Foundation (NSF) in 2006.

A whole cell computational model for the bacterium *Mycoplasma genitalium*, including all its 525 genes, gene products, and their interactions, was built by scientists from Stanford University and the J. Craig Venter Institute and published on 20 July 2012 in Cell.

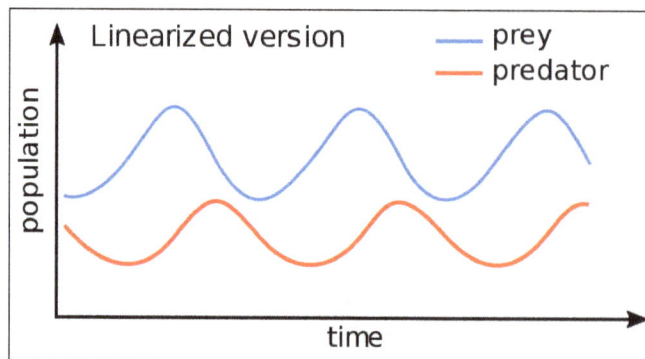

A sample time-series of the Lotka–Volterra model.
Note that the two populations exhibit cyclic behavior.

A dynamic computer model of intracellular signaling was the basis for Merrimack Pharmaceuticals to discover the target for their cancer medicine MM-111.

Membrane computing is the task of modelling specifically a cell membrane.

Multi-cellular Organism simulation

An open source simulation of C. elegans at the cellular level is being pursued by the OpenWorm community. So far the physics engine Gepetto has been built and models of the neural connectome and a muscle cell have been created in the NeuroML format.

Protein Folding

Protein structure prediction is the prediction of the three-dimensional structure of a protein from its amino acid sequence—that is, the prediction of a protein's tertiary structure from its primary structure. It is one of the most important goals pursued by bioinformatics and theoretical chemistry. Protein structure prediction is of high importance in medicine (for example, in drug design) and biotechnology (for example, in the design of novel enzymes). Every two years, the performance of current methods is assessed in the CASP experiment.

Human Biological Systems

Brain Model

The Blue Brain Project is an attempt to create a synthetic brain by reverse-engineering the mammalian brain down to the molecular level. The aim of this project, founded in May 2005 by the Brain and Mind Institute of the *École Polytechnique* in Lausanne, Switzerland, is to study the brain's architectural and functional principles. The project is headed by the Institute's director, Henry Markram. Using a Blue Gene supercomputer running Michael Hines's NEURON software, the simulation does not consist simply of an artificial neural network, but involves a partially biologically realistic model of neurons. It is hoped by its proponents that it will eventually shed light on the nature of consciousness. There are a number of sub-projects, including the Cajal Blue Brain, coordinated by the Supercomputing and Visualization Center of Madrid (CeSViMa), and others run by universities and independent laboratories in the UK, U.S., and Israel. The Human Brain Project builds on the work of the Blue Brain Project. It is one of six pilot projects in the Future Emerging Technologies Research Program of the European Commission, competing for a billion euro funding.

Model of the Immune System

The last decade has seen the emergence of a growing number of simulations of the immune system.

Virtual Liver

The Virtual Liver project is a 43 million euro research program funded by the German Government, made up of seventy research group distributed across Germany. The goal is to produce a virtual liver, a dynamic mathematical model that represents human liver physiology, morphology and function.

Tree Model

Electronic trees (e-trees) usually use L-systems to simulate growth. L-systems are very important in the field of complexity science and A-life. A universally accepted system for describing changes in plant morphology at the cellular or modular level has yet to be devised. The most widely implemented tree generating algorithms are described in the papers "Creation and Rendering of Realistic Trees", and Real-Time Tree Rendering.

Ecological Models

Ecosystem models are mathematical representations of ecosystems. Typically they simplify complex foodwebs down to their major components or trophic levels, and quantify these as either numbers of organisms, biomass or the inventory/concentration of some pertinent chemical element (for instance, carbon or a nutrient species such as nitrogen or phosphorus).

Models in Ecotoxicology

The purpose of models in ecotoxicology is the understanding, simulation and prediction of effects caused by toxicants in the environment. Most current models describe effects on one of many different levels of biological organization (e.g. organisms or populations). A challenge is the development of models that predict effects across biological scales. Ecotoxicology and models discusses some types of ecotoxicological models and provides links to many others.

Modelling of Infectious Disease

It is possible to model the progress of most infectious diseases mathematically to discover the likely outcome of an epidemic or to help manage them by vaccination. This field tries to find parameters for various infectious diseases and to use those parameters to make useful calculations about the effects of a mass vaccination programme.

Cellular Model

Creating a cellular model has been a particularly challenging task of systems biology and mathematical biology. It involves developing efficient algorithms, data structures, visualization and communication tools to orchestrate the integration of large quantities of biological data with the goal of computer modeling.

It is also directly associated with bioinformatics, computational biology and Artificial life.

It involves the use of computer simulations of the many cellular subsystems such as the networks of metabolites and enzymes which comprise metabolism, signal transduction pathways and gene regulatory networks to both analyze and visualize the complex connections of these cellular processes.

The complex network of biochemical reaction/transport processes and their spatial organization make the development of a predictive model of a living cell a grand challenge for the 21st century.

The eukaryotic cell cycle is very complex and is one of the most studied topics, since its misregulation leads to cancers. It is possibly a good example of a mathematical model as it deals with simple calculus but gives valid results. Two research groups have produced several models of the cell cycle simulating several organisms. They have recently produced a generic eukaryotic cell cycle model which can represent a particular eukaryote depending on the values of the parameters, demonstrating that the idiosyncrasies of the individual cell cycles are due to different protein concentrations and affinities, while the underlying mechanisms are conserved.

By means of a system of ordinary differential equations these models show the change in time (dynamical system) of the protein inside a single typical cell; this type of model is called a deterministic process (whereas a model describing a statistical distribution of protein concentrations in a population of cells is called a stochastic process).

To obtain these equations an iterative series of steps must be done: first the several models and observations are combined to form a consensus diagram and the appropriate kinetic laws are chosen to write the differential equations, such as rate kinetics for stoichiometric reactions, Michaelis-Menten kinetics for enzyme substrate reactions and Goldbeter–Koshland kinetics for ultrasensitive transcription factors, afterwards the parameters of the equations (rate constants, enzyme efficiency coefficients and Michaelis constants) must be fitted to match observations; when they cannot be fitted the kinetic equation is revised and when that is not possible the wiring diagram is modified. The parameters are fitted and validated using observations of both wild type and mutants, such as protein half-life and cell size.

In order to fit the parameters the differential equations need to be studied. This can be done either by simulation or by analysis.

In a simulation, given a starting vector (list of the values of the variables), the progression of the system is calculated by solving the equations at each time-frame in small increments.

In analysis, the properties of the equations are used to investigate the behavior of the system depending of the values of the parameters and variables. A system of differential equations can be represented as a vector field, where each vector described the change (in concentration of two or more protein) determining where and how fast the trajectory (simulation) is heading. Vector fields can have several special points: a stable point, called a sink, that attracts in all directions (forcing the concentrations to be at a certain value), an unstable point, either a source or a saddle point which repels (forcing the concentrations to change away from a certain value), and a limit cycle, a closed trajectory towards which several trajectories spiral towards (making the concentrations oscillate).

A better representation which can handle the large number of variables and parameters is called a bifurcation diagram (bifurcation theory): the presence of these special steady-state points at certain values of a parameter (e.g. mass) is represented by a point and once the parameter passes a certain value, a qualitative change occurs, called a bifurcation, in which the nature of the space changes, with profound consequences for the protein concentrations: the cell cycle has phases (partially corresponding to G1 and G2) in which mass, via a stable point, controls cyclin levels, and phases (S and M phases) in which the concentrations change independently, but once the phase has changed at a bifurcation event (cell cycle checkpoint), the system cannot go back to the previous levels since at the current mass the vector field is profoundly different and the mass cannot be reversed back through the bifurcation event, making a checkpoint irreversible. In particular the S and M checkpoints are regulated by means of special bifurcations called a Hopf bifurcation and an infinite period bifurcation.

BIFURCATION DIAGRAM

Molecular Level Simulations

Cell Collective is a modeling software that enables one to house dynamical biological data, build computational models, stimulate, break and recreate models. The development is led by Tomas Helikar, a researcher within the field of computational biology. It is designed for biologists, students learning about computational biology, teachers focused on teaching life sciences, and researchers within the field of life science. The complexities of math and computer science are built into the backend and one can learn about the methods used for modeling biological species, but complex math equations, algorithms, programming are not required and hence won't impede model building.

The mathematical framework behind Cell Collective is based on a common qualitative (discrete) modeling technique where the regulatory mechanism of each node is described with a logical function.

Model validation The model was constructed using local (e.g., protein–protein interaction) information from the primary literature. In other words, during the construction phase of the model, there was no attempt to determine the local interactions based on any other larger phenotypes or phenomena. However, after the model was completed, verification of the accuracy of the model involved testing it for the ability to reproduce complex input–output phenomena that have been observed in the laboratory. To do this, the T-cell model was simulated under a multitude of cellular conditions and analyzed in terms of input–output dose–response curves to determine whether the model behaves as expected, including various downstream effects as a result of activation of the TCR, G-protein-coupled receptor, cytokine, and integrin pathways.

The E-Cell Project aims "to make precise whole cell simulation at the molecular level possible".

CytoSolve - developed by V. A. Shiva Ayyadurai and C. Forbes Dewey Jr. of Department of Biological Engineering at the Massachusetts Institute of Technology - provided a method to model the whole cell by dynamically integrating multiple molecular pathway models.

In the July 2012 issue of Cell, a team led by Markus Covert at Stanford published the most complete computational model of a cell to date. The model of the roughly 500-gene Mycoplasma genitalium contains 28 algorithmically-independent components incorporating work from over 900 sources. It accounts for interactions of the complete genome, transcriptome, proteome, and metabolome of the organism, marking a significant advancement for the field.

Most attempts at modeling cell cycle processes have focused on the broad, complicated molecular interactions of many different chemicals, including several cyclin and cyclin-dependent kinase molecules as they correspond to the S, M, G1 and G2 phases of the cell cycle. In a 2014 published article in PLOS computational biology, collaborators at University of Oxford, Virginia Tech and Institut de Génétique et Développement de Rennes produced a simplified model of the cell cycle using only one cyclin/CDK interaction. This model showed the ability to control totally functional cell division through regulation and manipulation only the one interaction, and even allowed researchers to skip phases through varying the concentration of CDK. This model could help understand how the relatively simple interactions of one chemical translate to a cellular level model of cell division.

Simulated Growth of Plants

The simulated growth of plants is a significant task in of systems biology and mathematical biology, which seeks to reproduce plant morphology with computer software. Electronic trees (e-trees) usually use L-systems to simulate growth. L-systems are very important in the field of complexity science and A-life. A universally accepted system for describing changes in plant morphology at the cellular or modular level has yet to be devised.

'Weeds', generated using an L-system in 3D.

The realistic modeling of plant growth is of high value to biology, but also for computer games.

Theory + Algorithms

A biologist, Aristid Lindenmayer worked with yeast and filamentous fungi and studied the growth patterns of various types of algae, such as the blue/green bacteria *Anabaena catenula*. Originally the L-systems were devised to provide a formal description of the development of such simple multicellular organisms, and to illustrate the neighbourhood relationships between plant cells. Later on, this system was extended to describe higher plants and complex branching structures. Central to L-systems, is the notion of rewriting, where the basic idea is to define complex objects by successively replacing parts of a simple object using a set of rewriting rules or productions. The rewriting can be carried out recursively. L-Systems are also closely related to Koch curves.

Environmental Interaction

A challenge for plant simulations is to consistently integrate environmental factors, such as surrounding plants, obstructions, water and mineral availability, and lighting conditions. is to build virtual/environments with as many parameters as computationally feasible, thereby, not only simulating the growth of the plant, but also the environment it is growing within, and, in fact, whole ecosystems. Changes in resource availability influence plant growth, which in turn results in a change of resource availability. Powerful models and powerful hardware will be necessary to effectively simulate these recursive interactions of recursive structures.

Software

- OpenAlea: an open-source software environment for plant modeling, which contains L-Py, an open-source python implementation of the Lindenmayer systems.

- Branching: L-system Tree. A Java applet and its source code (open source) of the botanical tree growth simulation using the L-system.

- Arbaro- opensource.

- Treal- opensource.

- L-arbor.

- Genesis 3.0.

- AmapSim - from Cirad.

- GreenLab.

- ONETREE -Accompanying the CDROM is a CO_2 meter that plugs into your local serial port. It is this that controls the growth rate of the trees. It is the actual carbon dioxide level right at your computer that controls the growth rate of these virtual trees.

Ecosystem Model

An ecosystem model is an abstract, usually mathematical, representation of an ecological system (ranging in scale from an individual population, to an ecological community, or even an entire biome), which is studied to better understand the real system.

Using data gathered from the field, ecological relationships—such as the relation of sunlight and water availability to photosynthetic rate, or that between predator and prey populations—are derived, and these are combined to form ecosystem models. These model systems are then studied in order to make predictions about the dynamics of the real system. Often, the study of inaccuracies in the model (when compared to empirical observations) will lead to the generation of hypotheses about possible ecological relations that are not yet known or well understood. Models enable researchers to simulate large-scale experiments that would be too costly or unethical to perform on a real ecosystem. They also enable the simulation of ecological processes over very long periods of time (i.e. simulating a process that takes centuries in reality, can be done in a matter of minutes in a computer model).

Ecosystem models have applications in a wide variety of disciplines, such as natural resource management, ecotoxicology and environmental health, agriculture, and wildlife conservation. Ecological modelling has even been applied to archaeology with varying degrees of success, for example, combining with archaeological models to explain the diversity and mobility of stone tools.

Types of Models

There are two major types of ecological models, which are generally applied to different types of problems: (1) analytic models and (2) simulation / computational models. Analytic models are typically relatively simple (often linear) systems, that can be accurately described by a set of mathematical equations whose behavior is well-known. Simulation models on the other hand, use numerical techniques to solve problems for which analytic solutions are impractical or impossible. Simulation models tend to be more widely used, and are generally considered more ecologically realistic, while analytic models are valued for their mathematical elegance and explanatory power. Ecopath is a powerful software system which uses simulation and computational methods to model marine ecosystems. It is widely used by marine and fisheries scientists as a tool for modelling and visualising the complex relationships that exist in real world marine ecosystems.

Model Design

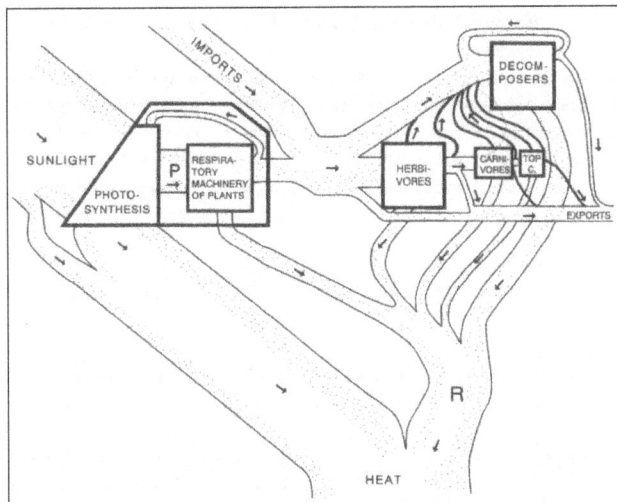

Diagram of the Silver Springs model (Odum, 1971). Note the aggregation into functional groups such as "herbivores" or "decomposers".

The process of model design begins with a specification of the problem to be solved, and the objectives for the model.

Ecological systems are composed of an enormous number of biotic and abiotic factors that interact with each other in ways that are often unpredictable, or so complex as to be impossible to incorporate into a computable model. Because of this complexity, ecosystem models typically simplify the systems they are studying to a limited number of components that are well understood, and deemed relevant to the problem that the model is intended to solve.

The process of simplification typically reduces an ecosystem to a small number of state variables and mathematical functions that describe the nature of the relationships between them. The number of ecosystem components that are incorporated into the model is limited by aggregating similar processes and entities into functional groups that are treated as a unit.

After establishing the components to be modeled and the relationships between them, another important factor in ecosystem model structure is the representation of space used. Historically, models have often ignored the confounding issue of space. However, for many ecological problems spatial dynamics are an important part of the problem, with different spatial environments leading to very different outcomes. *Spatially explicit models* (also called "spatially distributed" or "landscape" models) attempt to incorporate a heterogeneous spatial environment into the model. A spatial model is one that has one or more state variables that are a function of space, or can be related to other spatial variables.

Validation

After construction, models are *validated* to ensure that the results are acceptably accurate or realistic. One method is to test the model with multiple sets of data that are independent of the actual system being studied. This is important since certain inputs can cause a faulty model to output correct results. Another method of validation is to compare the model's output with data collected from field observations. Researchers frequently specify beforehand how much of a disparity they are willing to accept between parameters output by a model and those computed from field data.

The Lotka–Volterra Equations

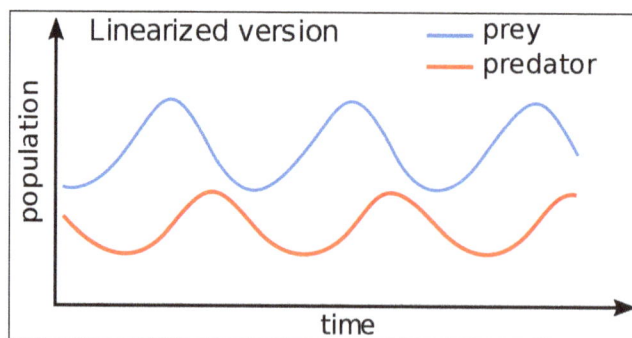

A sample time-series of the Lotka-Volterra model. Note that the two populations exhibit cyclic behavior, and that the predator cycle lags behind that of the prey.

One of the earliest, and most well-known, ecological models is the predator-prey model of Alfred J. Lotka and Vito Volterra. This model takes the form of a pair of ordinary differential equations, one representing a prey species, the other its predator.

$$\frac{dX}{dt} = \alpha.X - \beta.X.Y$$

$$\frac{dY}{dt} = \gamma.\beta.X.Y - \delta.Y$$

where,

- X is the number/concentration of the prey species;
- Y is the number/concentration of the predator species;
- α is the prey species' growth rate;
- β is the predation rate of Y upon X;
- γ is the assimilation efficiency of Y;
- δ is the mortality rate of the predator species.

Volterra originally devised the model to explain fluctuations in fish and shark populations observed in the Adriatic Sea after the First World War (when fishing was curtailed). However, the equations have subsequently been applied more generally. Although simple, they illustrate some of the salient features of ecological models: modelled biological populations experience growth, interact with other populations (as either predators, prey or competitors) and suffer mortality.

A credible, simple alternative to the Lotka-Volterra predator-prey model and its common prey dependent generalizations is the ratio dependent or Arditi-Ginzburg model. The two are the extremes of the spectrum of predator interference models. According to the authors of the alternative view, the data show that true interactions in nature are so far from the Lotka-Volterra extreme on the interference spectrum that the model can simply be discounted as wrong. They are much closer to the ratio dependent extreme, so if a simple model is needed one can use the Arditi-Ginzburg model as the first approximation.

Others:

The theoretical ecologist Robert Ulanowicz has used information theory tools to describe the structure of ecosystems, emphasizing mutual information (correlations) in studied systems. Drawing on this methodology and prior observations of complex ecosystems, Ulanowicz depicts approaches to determining the stress levels on ecosystems and predicting system reactions to defined types of alteration in their settings (such as increased or reduced energy flow, and eutrophication.

Conway's Game of Life and its variations model ecosystems where proximity of the members of a population are factors in population growth.

Protein Folding

Proteins have several layers of structure each of which is important in the process of protein folding. The first most basic level of this structure is the sequence of amino acids themselves. The sequencing is important because it will determine the types of interactions seen in the protein as it is folding. A novel sequence-based method based on the assumption that protein-protein interactions are more related to amino acids at the surface than those at the core.This study shows that not only is the amino acids that are in a protein important but also the order in which they are sequenced. The interactions of the amino acids will determine what the secondary and tertiary structure of the protein will be.

The next layer in protein structure is the secondary structure. The secondary structure includes architectural structures that extend in one dimension. Secondary structure includes α-Helixes and β-sheets. The α-helices, the most common secondary structure in proteins, the peptide –CO–NH– groups in the backbone form chains held together by NH ‾OC hydrogen bonds." The α-helices form the backbone of proteins and help to aid in the folding process. The β-sheets form in two distinct ways. They are able to form in both parallel β-pleated sheets and anti parallel β-pleated sheets.When the α-helix or β-sheet is formed, the excluded volumes generated by the backbone and side chains overlap, leading to an increase in the total volume available to the translational displacement of water molecules. This is important because it leads to a more thermodynamically stable conformation and leads to less strain on the protein as a whole and thus are aided by the conformation.

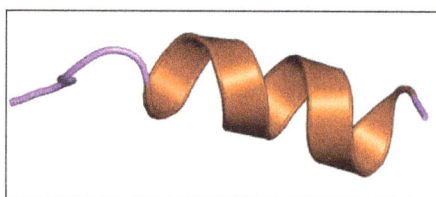

Typical example to an α-helix.   Typical example of an β-sheet.

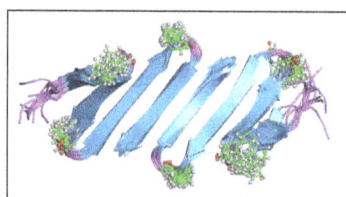

The tertiary structure is the next layer in protein structure. This takes the α-Helixes and β-sheets and allows them to fold into a three dimensional structure. Most proteins take on a globular structure once folded. The description of globular protein structures as an ensemble of contiguous closed loops or tightened end fragments reveals fold elements crucial for the formation of stable structures and for navigating the very process of protein folding. The globular proteins generally have a hydrophobic core surrounded by a hydrophilic outer layer. These interactions are important because they lead to the global structure and help create channels and binding sites for enzymes.

The last layer of protein structure is the quaternary structure. The folding transition and the functional transitions between useful states are encoded in the linear sequence of amino acids, and a long- term goal of structural biology is to be able to predict both the structure and function of molecules from the information in the sequence. The Subunit organization is the last level of structure in protein molecules. The organization of the subunits is important because that determines the types of interactions that can form and dictates its use in the body.

Protein Folding

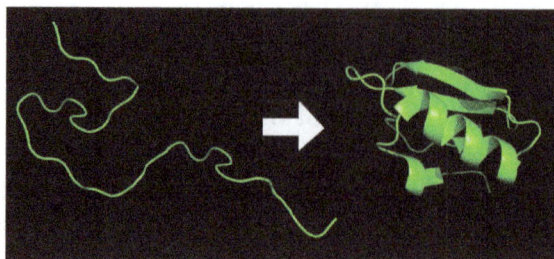

Protein Folding.

Proteins are folded and held together by several forms of molecular interactions. The molecular interactions include the thermodynamic stability of the complex, the hydrophobic interactions and the disulfide bonds formed in the proteins. The figure below is an example of protein folding.

The biggest factor in a proteins ability to fold is the thermodynamics of the structure. The interaction scheme includes the short-range propensity to form extended conformations, residue-dependent long-range contact potentials, and orientation-dependent hydrogen bonds. The thermodynamics are a main stabilizing force within a protein because if it is not in the lowest energy conformation it will continue to move and adjust until it finds its most stable state. The use of energy diagrams and maps are key in finding out when the protein is in the most stable form possible.

The next type of interaction in protein folding is the hydrophobic interactions within the protein. The framework model and the hydrophobic collapse model represent two canonical descriptions of the protein folding process. The first places primary reliance on the short-range interactions of secondary structure and the second assigns greater importance to the long-range interactions of tertiary structure. These hydrophobic interactions have an impact not just on the primary structure but then lead to changes seen in the secondary and tertiary structure as well. Globular proteins acquire distinct compact native con- formations in water as a result of the hydrophobic effect. When a protein has been folded in the correct way it usually exists with the hydrophobic core as a result of being hydrated by waters in the system around it which is important because it creates a charged core to the protein and can lead to the creation of channels within the protein. The hydrophobic interactions are found to affect time correlation functions in the vicinity of the native state even though they have no impact on same time characteristics of the structure fluctuations around the native state. The hydrophobic interactions are shown to have an impact on the protein even after it has found the most stable conformation in how the proteins can interact with each other as well as folding themselves.

Another type of interaction seen when the protein is folding is the disulfide linkages that form in the protein. The disulfide bond, a sulfur- sulfur chemical bond that results from an oxidative process that links nonadjacent (in most cases) cysteine's of a protein. These are a major way that proteins get into their folded form. The types of disulfide bonds are cysteine-cysteine linkage is a stable part of their final folded structure and those in which pairs of cysteines alternate between the reduced and oxidized states. The more common is the linkages that cause the protein to fold together and link back on itself compared to the cysteines that are changing oxidation states because the bonds between cysteines once created are fairly stable.

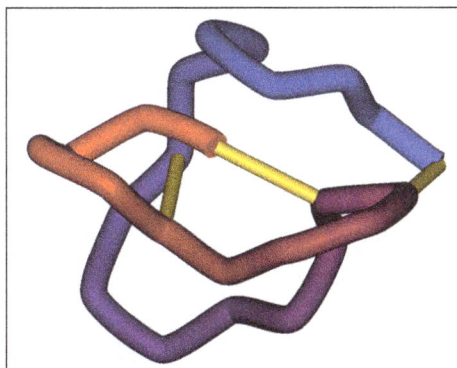

Disulfide Bonds, shown in the picture in yellow.

Misfunctions

Proteins can miss function for several reasons. When a protein is miss folded it can lead to denaturation of the protein. Denaturation is the loss of protein structure and function. The miss folding does not always lead to complete lack of function but only partial loss of functionality. The miss functioning of proteins can sometimes lead to diseases in the human body.

Alzheimer's Disease

Alzheimer's Disease (AD) is a neurological degenerative disease that affects around 5 million Americans, including nearly half of those who are age 85 or older. The predominant risk factors of AD are age, family history, and heredity. Alzheimer's disease typically results in memory loss, confusion of time and place, misplacing places, and changes in mood and behavior. AD results in dense plaques in the brain that are comprised of fibrillar β-amyloid proteins with a well-orders β-sheet secondary structure. These plaques visually look like voids in the brain matter and are directly connected to the deterioration of thought processes. It has been determined that AD is a protein misfolding disease, where the misfolded protein is directly related to the formation of these plaques in the brain.

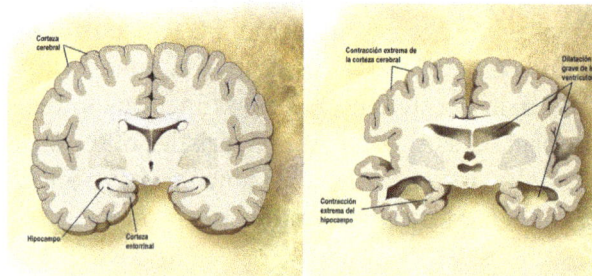

Comparison of healthy brain (left) with brian with Alzheimer's (right).

It is yet to be fully understood what exactly causes this protein misfolding to begin, but several theories point to oxidative stress in the brain to be the initiating factor. This oxidation results in damage to the phospholipids in the brain, which has been found to result in a faster accumulation of amyloid β-proteins.

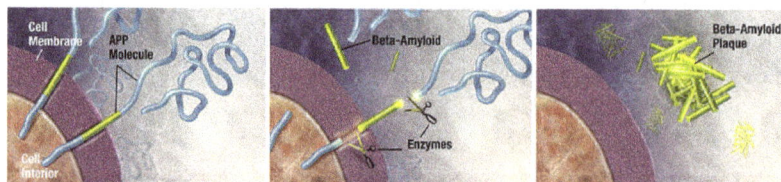

Beta-Amyloid Plaque Formation.

Cystic Fibrosis

Cystic Fibrosis (CF) is a chronic disease that affects 30,000 Americans. The typical affects of CF is a production of thick, sticky mucus that clogs the lungs and leads to life-threatening lung infection, and obstructs the pancreas preventing proper food processing. CF is caused by protein misfolding. This misfolding then results in some change in the protein known as cystic fibrosis transmembrane conductance regulator (CFTR), which can result in this potentially fatal disease. In approximately

70% of CF cases, a deletion of phenylalanine at position 508 in the CFTR is deleted. This deletion of Phe508 seems to be directly connected to the formation of CF. The protein misfolding that results in CF occurs prior to birth, but it is not entirely clear as to why.

References

- Millspaugh, Joshua J.; et al. (2008). "General Principles for Developing Landscape Models for Wildlife Conservation". Models for planning wildlife conservation in large landscapes. Academic Press. P. 1. ISBN 978-0-12-373631-4

- Systems-biology, science: britannica.com, Retrieved 26 July, 2019

- Biograf 3R - Computational Alternatives to Animal Testing - Home". Biograf.ch. Doi:10.1016/j.taap.2012.03.018. Retrieved 2013-05-22

- Biological, network, teaching, francesc: sci.unich.it, Retrieved 21 May, 2019

- Al-Lazikani, Bissan; Banerji, Udai; Workman, Paul (2012). "Combinatorial drug therapy for cancer in the post-genomic era". Nature Biotechnology. 30 (7): 679–692. Doi:10.1038/nbt.2284. PMID 22781697

- Dale, Virginia H. (2003). "Opportunities for Using Ecological Models for Resource Management". Ecological Modeling for Resource Management. Pp. 3–19. Doi:10.1007/0-387-21563-8-1. ISBN 978-0-387-95493-6

- Protein-Folding, Protein-Structure, Proteins, Supplemental-Modules-(Biological-Chemistry), Biological-Chemistry: libretexts.org, Retrieved 8 January, 2019

Permissions

All chapters in this book are published with permission under the Creative Commons Attribution Share Alike License or equivalent. Every chapter published in this book has been scrutinized by our experts. Their significance has been extensively debated. The topics covered herein carry significant information for a comprehensive understanding. They may even be implemented as practical applications or may be referred to as a beginning point for further studies.

We would like to thank the editorial team for lending their expertise to make the book truly unique. They have played a crucial role in the development of this book. Without their invaluable contributions this book wouldn't have been possible. They have made vital efforts to compile up to date information on the varied aspects of this subject to make this book a valuable addition to the collection of many professionals and students.

This book was conceptualized with the vision of imparting up-to-date and integrated information in this field. To ensure the same, a matchless editorial board was set up. Every individual on the board went through rigorous rounds of assessment to prove their worth. After which they invested a large part of their time researching and compiling the most relevant data for our readers.

The editorial board has been involved in producing this book since its inception. They have spent rigorous hours researching and exploring the diverse topics which have resulted in the successful publishing of this book. They have passed on their knowledge of decades through this book. To expedite this challenging task, the publisher supported the team at every step. A small team of assistant editors was also appointed to further simplify the editing procedure and attain best results for the readers.

Apart from the editorial board, the designing team has also invested a significant amount of their time in understanding the subject and creating the most relevant covers. They scrutinized every image to scout for the most suitable representation of the subject and create an appropriate cover for the book.

The publishing team has been an ardent support to the editorial, designing and production team. Their endless efforts to recruit the best for this project, has resulted in the accomplishment of this book. They are a veteran in the field of academics and their pool of knowledge is as vast as their experience in printing. Their expertise and guidance has proved useful at every step. Their uncompromising quality standards have made this book an exceptional effort. Their encouragement from time to time has been an inspiration for everyone.

The publisher and the editorial board hope that this book will prove to be a valuable piece of knowledge for students, practitioners and scholars across the globe.

Index

www.ingramcontent.com/pod-product-compliance
Lightning Source LLC
Chambersburg PA
CBHW082029190326
41458CB00010B/3313